A HANDBOOK OF DAIRY FOODS

CONTENTS

Acknowledgements

The Council is indebted to the Department of Experimental Medicine, Cambridge, for their help in compiling the tables showing the nutritional composition of dairy foods, as well as to all those people in educational, medical, Government and dairy industry circles who have assisted in the preparation of the final manuscript.

Chapter I

HISTORICAL BACKGROUND

Cattle Breeding

Historians have recorded various developments in farming and food production in this country throughout many early centuries. However, because the open-fields system of agriculture prevailed until the seventeenth century, little control could be exercised. The situation changed during the reign of *George III*, known as *"Farmer George"*, (1760-1820) when enclosure of land encouraged greater food production and gave farmers an opportunity to improve their cattle by controlled selective breeding.

Improvement of Livestock

One of the major problems of livestock farming was winter feeding and the majority of the cattle had to be slaughtered and salted in the autumn irrespective of their value. Improvements in this situation were brought about by the efforts of *Lord "Turnip" Townshend* (1674-1738) and *Richard Weston* (1735-1806). The former introduced turnips and the latter clover for feeding cattle through the winter. Thus valuable animals could be saved and used for breeding over a number of years.

Among those who pioneered the improvement of livestock, *Robert Bakewell* (1725-1795) stands out. At a time when the demand for meat-producing cattle was increasing, he went to endless trouble studying the good and bad points of his animals and recording these for future reference.

Arthur Young (appointed Secretary to the newly-formed Board of Agriculture* in 1793), generated enthusiasm and spread news of the new ideas and developments of the more up-to-date farmers.

One of the first improvers of dairy cattle was *Thomas Bates*, a yeoman farmer who lived on Tees-side. In 1839 he exhibited Shorthorn cattle at the first show organized by the English Agricultural Society.† These had high milk yields.

Herd Books and Breed Societies

A major development in scientific breeding practice was the institution of *herd books*. The earliest of these was *Coates's Herd Book for Improved Shorthorn Cattle*, first published in 1822.

Herd books recorded pedigrees of the best breeding animals in the country. With the subsequent formation of breed societies to keep up and issue herd books, standards of quality were established which breeders could attain and through which the best characteristics could be preserved for controlled breeding from selected and approved types of cattle. By 1914, all of the leading breeds were represented by Herd Book Societies.

Breeding to type was also encouraged by the growth throughout the

* *Now the Ministry of Agriculture, Fisheries and Food*
† *Now the Royal Agricultural Society of England*

nineteenth century of agricultural and dairy shows holding milking trials and making awards. Alongside this, there was increased interest in the scientific theory of animal breeding and from the latter part of the nineteenth century onwards several important pioneering studies were published on heredity, genetics, animal fertility and reproduction (by *Galton, Ewart, Heape, Bateson* and others).

In the years just before the First World War, *James Speir* of Newton, Ayrshire, developed the practice of recording the milk yield of cows. In more recent times the introduction of artificial insemination has revolutionized cattle breeding (see page 9).

Milking Machines

The first milking machine (1849) was an American invention. It worked on the siphon principle but was found to be injurious, so that later inventors based their experiments on a suction action. The first suction machine, also American, was exhibited in this country in 1862. It drew milk from the cow at a rate of a gallon (4.55 litres) a minute. The patent was sold for £5,000.

From 1878 onwards several designs seeking to imitate hand milking were put forward, but all failed because it was impossible to keep them clean. In 1889 the first British continuous suction machine was invented by *William Murchland*, but in 1892 the principle of operation was changed to intermittent suction as the early machines were damaging the cows' teats.

Dr. Shields of Glasgow took out patents for his 'pulsator' in 1895-96 and other developments followed including the important *Gillies teat-cup* (1902). Early in the 1900s machines incorporating all the principles of the modern mechanical milker were introduced.

Milk Supplies

Before the 1850s clean milk was almost unknown and the consumption of liquid milk was low. Many cows were diseased; cowsheds were often dirty; and so were utensils which provided a perfect breeding-ground for germs. Milk would often go sour in a few hours during hot weather so that much of it was turned into butter and cheese, but even this was often rancid when sold. In these unhygienic conditions formaldehyde or borax was frequently added in an effort to combat the dangers of diseases.

Town Milk

Milk for the smaller towns was supplied by farms on the outskirts, but it usually arrived in a dirty or sour condition heavily adulterated with water (amongst other things). It was often carried through the dusty and dirty streets in open pails.

Most city milk was supplied by the town herds and cows were often kept at the rear of the milk shops while some were milked in the streets. Some of the town herds were large and cows for them purchased in calf, usually from the neighbouring counties. There are accounts of cows being driven from Wales to London, while Holland too supplied heavy milking cows to some English town herds.

Conditions in the town cowsheds, which were often quite large, were cramped and too many cattle were housed in very unhygienic buildings. The nearest approach to a public health authority was the *Inspector of Nuisances* who among many duties was entitled to order that "any accumulation of dung, soil, filth or other obnoxious matter shall be removed by the person to whom the same belongs".

Cleaner Milk

Gradually concern for cleanliness and the care of milk began to appear. A large herd in Kensington consisted of pedigree Shorthorns. It was evidently an outstanding one producing nearly 300 gallons (1,350 litres) of milk daily. The yield was carefully recorded and an attempt made to keep the milk in a cool condition, the premises being "admirably cool and sweet". But dairies with such a high standard of cleanliness as this were rare.

In Scotland a pioneer of clean milk production was *William Harley* of Glasgow. In 1809 he opened a dairy in George Square and insisted on cleanliness in the production of milk. He washed the cows before milking, and cleaned and steamed his utensils both before and after use. He made clean, sweet milk available to citizens of Glasgow for the first time; hitherto they had known nothing but sour milk. What is more he managed to reduce the price very considerably.

Developments in the 1860s

At the beginning of the 1860s there was a rise in the population and an improvement in living standards. These led to an increase in the demand for milk and had some influence on the expansion in dairy farming.

At the same time British farmers were beginning to find difficulty in competing with foreign prices for beef and wheat. They turned to milk production and began to build up pedigree dairy herds, and the *flying herd system* became common on big farms. This system involves buying cows in milk; keeping them for a time as milkers, then selling them for beef. It has the advantage that the herd is permanently in milk.

Cattle Plague

Outbreaks of cattle plague and rinderpest (which has by the present day been eliminated) recurred at intervals, while other diseases, especially pleuro-pneumonia and foot and mouth disease, were almost endemic. It was the advantage of the town dairies in producing fresh, if unhygienic, milk with the avoidance of transport costs which had kept them in business despite the difficulties. But in 1865, 1866 and 1867 there were unusually severe outbreaks of cattle plague which reduced the number of cattle in London very substantially.

These outbreaks led to the Government taking the first action ever to attempt to control cattle disease. The *Cattle Diseases Prevention Act*, 1866, introduced a slaughter policy for cattle plague which eventually was successful in eliminating the disease. The important thing, however, is that the drop in supplies of milk, which was brought about by these outbreaks of cattle disease, caused the dairymen to look elsewhere for their milk supplies. They were helped in their search by the substantially increased mileage of railways which were available by this time.

Railways

In the 1860s, with the development of the railways, the long-distance haulage of milk really began. Great Western Railway records, published in various places, show that in the month of January 1865, before the full severity of the outbreaks of cattle plague had been felt, they carried just under 9,000 gallons (40,000 litres) of milk. In January the following year this had gone up to 144,000 gallons (654,600 litres) and from then on milk haulage by railways increased substantially throughout the nineteenth century until, in 1900, the Great Western Railway alone was carrying something like 25 million gallons (113,650,000 litres) a year, chiefly from the West Country. In other words, they were carrying the produce of upwards of 50,000 cows from the West Country to London. It can be seen, therefore, that in the early stages the dispersal of milk production away from the towns into the countryside was closely bound up with the development of rail transport. One of the pioneers of railing milk from country to town was *Sir George Barham*. He started in 1866 and, within two years, more than 4 million gallons (over 18 million litres) of milk per year were brought to London in this way for him.

Mechanization

The first factory for manufacturing dairy foods was established in Derby in 1870 and by 1900 the factory system had spread throughout the country. New mechanical processes favoured factory production of butter, condensed milk and milk powder.

The introduction of the centrifugal separator in 1877 brought about a revolution in the production of cream. Cheese-making, too, was mechanized and many American methods were adopted in factories in the United Kingdom.

Cooling and refrigeration were somewhat neglected until the middle of the nineteenth century. The refrigeration apparatus in common use throughout the 1880s and 1890s was *Lawrence's cooler and aerator* — the forerunner of the surface cooler, which closely resembles it.

Bottling and Heat Treatment

From mid-Victorian times milk was delivered by hand or horse-drawn carts. Until well into the twentieth century it was generally sold loose — the milk-men filled the purchaser's jug from a churn or hand can in the street.

Milk was first bottled in the United Kingdom in 1884. The bottles were sealed with a wired-on closure, but unfortunately this first attempt to provide the public with cleaner milk ended in failure.

In 1887 *Anthony Hailwood*, a North Country dairyman, exhibited bottles of filtered, medicated and other milk at the Jubilee Exhibition in Manchester. This proved to be more successful and in 1894 he applied a method of sterilizing milk by heating it to a high temperature in swing-stoppered bottles similar to the old-style mineral water bottles.

Sterilized milk was one of the earliest products of enterprising dairymen. However, it must have been very unpleasant until the invention of the *homogenizer* by M. Gaulin in 1904 brought about an improvement.

A steady demand for this type of milk developed, although it was, and remains, localized in centres such as Birmingham, Manchester, Leeds and, to a lesser extent, London.

Pasteurization

Long before the experiments of *Louis Pasteur* (1822-1895) it was known that the application of heat to certain foodstuffs helped to conserve them. In 1795 *Nicolas Appert*, who invented canning, applied the process to milk. In the early part of his career, Pasteur investigated the problem of bacterial growth in milk and proved that souring was due to the multiplication of bacteria which, he believed, came from the atmosphere. He showed that the application of heat to milk would destroy many of the organisms and delay souring.

Many laboratory experiments took place between 1900 and 1920 to establish the combinations of temperature and time of heating required to destroy all milk-borne pathogenic organisms. Ever since the late nineteenth century milk heat treated by the Pasteur method had been sold by progressive dairy firms, but the authorities mistrusted them and delayed the introduction of legislation to define the pasteurization process.

In 1922 a British company invented the first *plate heat exchanger* and, at the same time, a completely reliable *holder process* plant was introduced. These changed the situation and legislation followed when the *Milk (Special Designations) Order* 1922, prescribed the conditions under which milk sold as 'pasteurized' had to be heat treated. This order required milk to be pasteurized by heating to between 62·8°C—65·8°C for 30 minutes, but it is now usually heated to 72°C for only 15 seconds.

Years of Consolidation

Apart from the introduction of pasteurization and the increasing use of bottles for pre-packing milk, the early years of the twentieth century were uneventful. These were years of consolidation and of increasing capitalization in distributing and processing. Even before the 1914-1918 war many of the wholesalers had combined to form powerful groups and the war accentuated the trend.

An important feature of the war years was the growing dissatisfaction with the quality of milk supply. Grading systems had already been in operation in the main cities of the United States and the distribution of graded milk in bottles was already well known there.

The Milk (Special Designations) Order, 1922, prescribed those designations, in addition to 'pasteurized', under which milk might be sold. Scottish legislation followed similar lines. In Northern Ireland pasteurization of milk was first regulated in 1934.

Although some pre-1900 efforts to improve the cleanliness of milk have been noted, these were somewhat localized in character. The turn of the century saw the beginning of two 'clean milk' campaigns which were to have a far-reaching effect. One was instigated at the National Institute for Research in Dairying and the other by the National Clean Milk Society.

National Institute for Research in Dairying

R. Stenhouse Williams joined the staff of the National Institute for Research in Dairying when it began modestly in 1912 with a staff of two — a bacteriologist and a chemist. He was in charge of the Institute and the Bacteriological Department when the Institute moved to Shinfield, near Reading, in 1923.

Dr. Stenhouse Williams was a medical man who had been a lecturer in bacteriology at Liverpool University and as such was attached to the hospitals there. The slum conditions in Liverpool had made him aware of the desperate need of a pure food, such as milk, which could supply a highly nutritious addition to the diet of poor families. He also knew that vast quantities of milk were wasted through being too sour to drink and that much of the remainder was heavily contaminated. Contaminated milk was killing thousands of young children every year, as well as undermining the health of many more.

One of the greatest dangers was *bovine tuberculosis.* Over 40 per cent of the national herd was infected, so Dr. Williams turned his attention to the farms. He toured the country in an effort to convince dairy farmers that clean milk was essential and that it also represented a financial saving as milk kept longer. He demonstrated good techniques and encouraged the use of covered pails and suitable clothes.

He proved that good-quality milk could be produced even in the hottest weather provided that the cows and milkers were clean and the equipment washed and sterilized.

National Clean Milk Society

Wilfred Buckley of Moundsmeer Farm, Hampshire, was Director of Milk Supplies for the Ministry of Agriculture from 1917—1920. In 1915 he founded the *National Clean Milk Society*, being its Chairman until 1928; with other volunteer members of the Society he campaigned vigorously for stricter control over producers and retailers. He also agitated for the introduction of a grading system, so that dairy farmers would be paid a higher price for better quality milk. The Society aroused so much interest that in 1915 its recommendations were placed on the Statute Book, but, mainly because of the war, the act was not brought into operation until 1923.

Other Developments

There were three other outstanding developments in the inter-war years. First, the improvements in road transport, second a new emphasis on education and nutrition research, and third the development of controlled marketing.

Just as the coming of the railways had revolutionized farm practice in the 1860s and 1870s, so the opening up of the farm butter and cheese-making areas by motor lorry brought about a further expansion in milk production and also new marketing problems. Milk was assembled at depots in the country by road and brought into the towns by road and rail — much of it in tankers.

Education and Nutrition Research

The second development was the new emphasis on human diet and nutrition. *The National Milk Publicity Council of England and Wales* was formed in 1920 and a campaign was started to raise the level of milk consumption in the country.

In 1927 the Council introduced a *school milk scheme* whereby ⅓ pint (190ml) of milk could be obtained for 1d. (just under ½p) per bottle including straws.

By 1933 nearly one million children were taking advantage of it and, under the *Milk Act of* 1934, operation of the scheme was taken over by the newly established Milk Marketing Board. Government subsidies made it possible to supply ⅓ pint of milk for ½d. to all children in elementary schools—those from poor homes obtained it free of charge. By 1939 more than half the children in state-aided schools were receiving cheap milk.

Other social welfare schemes of this period were subsidized by the rates and by the Treasury, and included the provision of cheap milk to mothers and infants in need.

Experimental Work

The above schemes were developed largely as a result of experimental work done by the National Milk Publicity Council and by *Dr. H. C. Corry Mann* and *Dr. J. Boyd Orr* (later Lord Boyd Orr) who carried out tests in the 1920s to investigate the dietetic value of milk and milk products. These tests were spread over several years and involved groups of school children in different parts of the United Kingdom.

In 1922 the National Milk Publicity Council started an experiment with the co-operation of the Birmingham Education Authorities whereby 50 under-nourished children were supplied with a pint of milk a day for a period of two months. A report on the experiment was published and aroused much interest. It was shown that the children who had been given milk had gained in height and weight far more than similar children without the additional milk. The report was circulated to Medical Officers of Health and other towns became interested in similar experiments.

In the subsequent Corry Mann experiment, boys (whose ages varied from six to eleven years) in a large boarding school were divided into seven groups and each group was given a different diet. Accurate observations concerning the boys' height, weight and general health were kept during the period of twelve months.

Group 1 had an ordinary diet with no extras. This was the control group.
Group 2 had the ordinary diet *plus* one pint of milk daily.
Group 3 had the ordinary diet *plus* extra sugar.
Group 4 had the ordinary diet *plus* extra butter.
Group 5 had the ordinary diet *plus* extra margarine.
Group 6 had the ordinary diet *plus* extra casein.
Group 7 had the ordinary diet *plus* extra watercress.

At the conclusion of this experiment it was found that the boys in Group 1 had only gained an average of 3lb (1·4kg) and grown an average of 1¼in. (32mm) whilst the boys in Group 2 had gained an average of 7lb (3·2kg) and grown an average of 2⅝in. (67mm). The rest of the groups varied between these two extremes, those with extra butter being the second highest. It was also noticed that the boys in Group 2 improved in both health and spirits.

A later experiment was carried out by the Scottish Board of Health under the supervision of Dr. Boyd Orr. The results confirmed those obtained in the Corry Mann experiment. Of the Scottish experiment 1,400 boys (whose ages were between five and fourteen years) in local country schools were divided into four groups.

Group 1 had an ordinary home diet.

Group 2 had the ordinary home diet *plus* extra milk.

Group 3 had the ordinary home diet *plus* separated milk.

Group 4 had the ordinary home diet *plus* one biscuit which had the equivalent energy value of the separated milk.

After seven months the group which had been receiving the extra milk gained an average of 20 per cent in height and weight over the others.

Another large-scale experiment was carried out between early 1935 and mid-1936. Some 8,000 children from Burton-on-Trent, Huddersfield, Luton, Wolverhampton and Renfrewshire took part and were divided into four groups.

Group 1 acted as 'controls' and received biscuits.

Group 2 received $\frac{1}{3}$ pint pasteurized milk.

Group 3 received $\frac{2}{3}$ pint pasteurized milk.

Group 4 received $\frac{2}{3}$ pint raw milk.

The children who had received $\frac{1}{3}$ pint milk showed greater improvement in general health, physique and scholastic ability than those who received biscuits. The improvement was greater among those who received $\frac{2}{3}$ pint milk, but there was no significant difference between those who had pasteurized milk and those who had raw milk.

In 1953 two Liverpool doctors of medicine, R. Hecker and H. H. Andrews, lived for a whole month on nothing but milk. During this time they carried out their normal work and one of them played championship tennis regularly. Dr. Andrews drank twelve pints (6·8 litres) of milk a day and Dr. Hecker ten pints (5·7 litres) a day.

Controlled Marketing

The third outstanding feature of this period was the development of controlled marketing. After several attempts by the producers and retailers to regulate prices by collective bargaining, the milk producers vested all control of prices and conditions of supply in their own elected *Milk Marketing Boards* (see page 16) who henceforth negotiated with buyers. At the outbreak of war in 1939 the Government immediately controlled milk prices and later took over responsibility for the wartime distribution and allocation of milk.

Wartime policies imposed severe restriction on the manufacture of cheese and butter, the sale of cream was banned and greatly increased quantities of milk became available for the liquid market. Milk production increased greatly but not as much as the demand and the adoption of a modified form of rationing for liquid milk became necessary between 1941 and 1950.

Welfare Schemes

The complete control of national milk supplies during the war was followed by the introduction of a system of milk priorities. *The National Milk Scheme** was introduced in July 1940 to provide milk free, or at a reduced price, for *all* expectant and nursing mothers and children under school age. In April 1971, cheap welfare milk was withdrawn and some additional categories were introduced into the issue of free welfare milk.

**Later this became part of the Welfare Foods Service*

To safeguard children's health, the *Milk-in-Schools Scheme* was extended during and after the war to private as well as State Schools. The scheme later operated under the Education Act of 1944 when all children attending school became entitled to an allowance of milk free of charge. In 1968 new legislation restricted the scheme to children attending state primary schools and a further Act in 1970 entitled junior pupils in middle schools to free milk. From 1971 free school milk was discontinued to pupils at the end of the summer term after they reached the age of seven. During 1977 the EEC allocated funds to partially subsidize school milk schemes in member countries and in 1978 the Government allowed local authorities the discretion to again provide free milk for junior pupils in maintained schools. The EEC regulations have now been extended to enable local authorities also to claim subsidies on milk and certain milk products provided with subsidized school meals.

In Northern Ireland, by statutory requirement, all children in primary schools serving meals receive ⅓ pint of milk per day and all children in primary schools which do not serve meals receive ⅔ pint per day.

Post War to the Present Day

Artificial Insemination

Dairy herd improvement has been greatly facilitated in recent years by the practice of artificial insemination, one of the most revolutionary developments in scientific breeding history. Though its use on a commercial scale in Britain is little more than 30 years old, the practice is now widely used in all areas of the country and for all types of herd. The principal advantage of A.I. is that one bull can sire well over 20,000 calves in a year. This compares with the pre-A.I. years when bulls sired on average about 50 calves a year by natural mating. By making the service of high-quality bulls readily and economically available to all farmers, A.I. has become a vital factor in upgrading national milk production.

Designated Milk

Since 1951 the areas in which only specially designated milk may normally be sold by retail have been progressively extended until by the middle of 1962 they covered the whole mainland of England and Wales. The specially designated milks at present are *Untreated* (which is raw), *Pasteurized*, *Sterilized,* and *Ultra Heat Treated*. In Scotland the designations are slightly different.

Production and Consumption Today

Over 13,000 million litres of milk are now produced in the United Kingdom each year and the average consumption of milk is about 134 litres per annum. This consumption is about four times what it was a century ago.

This increase reflects the improvements in standards during the last hundred years, and the main measures which have contributed to the safety of our milk supplies have been the decrease in the pollution of water supplies and the improved methods of bacteriological control; cleaner milking; improvements in the production, handling and distribution of milk to obtain a product of higher nutritional and hygienic quality; and the effectiveness of heat treatment processes — especially pasteurization.

Chapter II

MILK PRODUCTION

The Nutrition of Dairy Cows

Milk is made from the nutrients in the food which a cow eats and the water she drinks. Not all she eats goes to produce milk however; part of her food is for her own bodily needs. Certain quantities of energy-giving foods (carbohydrates etc) and proteins are required together with essential vitamins and minerals. The food from which the cow obtains these nutrients will vary according to the time of year, because of availability.

Grazing

In spring when the weather conditions are suitable and the grass has begun to grow, cows will be turned out of their winter quarters to graze. For most of the grazing season, the fresh grass will provide all the nutrients the cow requires unless it is a cow which produces a very high milk yield, in which case additional food in the form of high protein concentrate is often fed. When grazing a cow may eat up to 150 lb (70 kg) of grass a day.

The quality of the grass is therefore an important factor in milk production. Several factors decide the food value of the grass and meadow plants.

(a) Fertility of the soil — under natural conditions the fertility of the soil depends on the type of rock from which it is derived, and on the decomposing action of bacteria and other forms of microscopic life on dead animal and vegetable matter. When land is farmed however, some of these materials are removed and must be replaced by the use of farmyard manure and chemical fertilisers. In addition, because bacteria need air and good working conditions, it is important to keep the soil porous and well drained and to add lime as well in some cases.

(b) The quality of the grass — this varies with the time of year. Fresh spring grass is better than autumn grass and a young plant is much more digestible than one that is at the 'seed' stage.

(c) The variety of grass — some varieties will produce much greater quantities of high value grass. Modern plant breeding techniques have also produced superior strains which are more digestible and nutritious than others.

(d) Grazing systems — modern systems using electric fences or small paddocks enable small areas of grass to be grazed very rapidly and intensively for a short period and then 'rested' for a long period to grow again. This is much better than a constant grazing system as the plant is allowed to recover fully rather than continually being eaten off just as it starts to grow.

Winter Feeding

The main basis for the winter diet is usually grass which has been conserved by the farmer during the summer months, when it is plentiful. This takes the form of hay (sun-dried grass) or silage ('pickled' grass), and it is always less nutritious than the fresh grass from which it was made. Because of this, it is usually necessary to feed additional foods (these are compound foods containing grain, soya bean, earthnut or other high protein food) and this 'concentrate' is usually 'rationed' to the cow while she is being milked.

Other foods can be fed as an alternative to part of the diet, e.g. sugar beet pulp, reject potatoes, brewers' grains, while kale and cabbage are often especially grown for winter feeding.

Factors Affecting Milk Yield and Milk Quality

1. Breed—some breeds are noted for producing large quantities of milk (e.g. the Friesian) and some for high butterfat content (e.g. the Guernsey, Jersey and South Devon)
2. Quality and quantity of food
3. Management—good animal husbandry and efficient milking
4. Breed improvement programmes—the use of artificial insemination techniques enables farmers to use the best bulls in the country to improve the genetic milking ability of his herd.

Digestion in a Cow

A cow tears off her food or takes it from a manger and quickly swallows it. It is then stored in a special part of the digestive system so that she can masticate and digest it later at her leisure.

The cow's stomach is in four parts (see diagram below, this page). The food is stored in the first two of these—the *rumen* and the *reticulum*. The unchewed food (cud or bolus), soaked with saliva, is regurgitated from time to time in small amounts from the rumen to the mouth. Here the cow masticates it by chewing.

The masticated food returns to the rumen where it is fermented by micro-organisms which help the breakdown process and facilitate absorption by the cow. Some of the nutrients containing energy required by the cow are absorbed directly through the rumen wall into the blood stream.

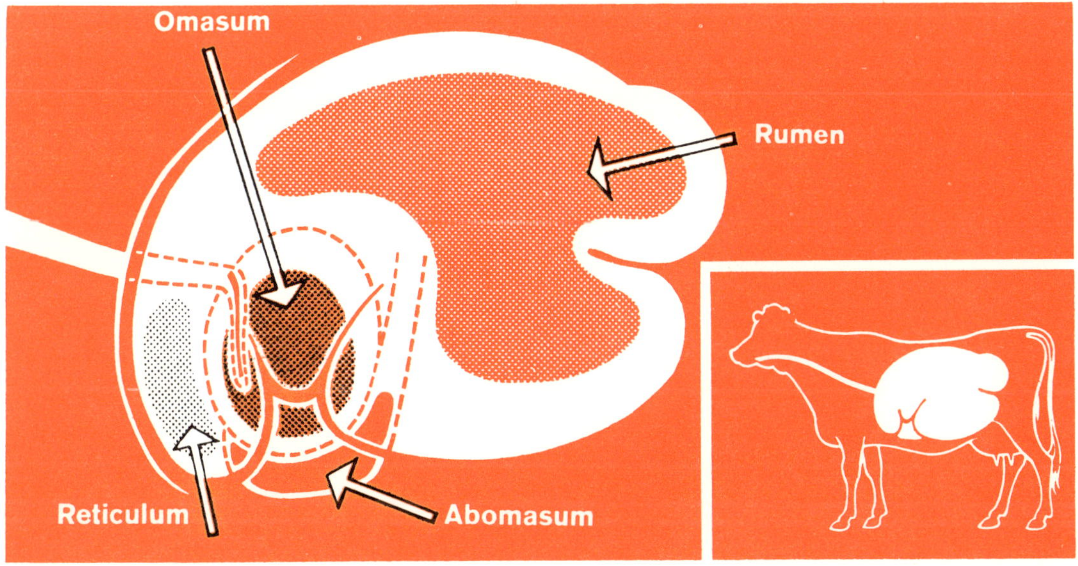

From Stomach to Udder: Function of Blood

The food mass passes through the *omasum* where surplus water, so necessary to move the food mass through the reticulo-rumen, is extracted. Eventually the food mass reaches the *abomasum* which has the functions of a true stomach. Here digestive juices are secreted and true chemical enzyme digestion begins. The food mass passes through the abomasum into the intestines where further absorption of nutrients takes place. The various nutrients absorbed in the stomach and intestines finally reach the blood, which circulates throughout the whole body to carry all the materials required to feed the body cells, maintain the body in good condition, and produce milk.

In the Udder

In the udder the blood passes through very small blood vessels surrounding groups of secretory cells known as *alveoli*. Milk is synthesized by the cells of each alveolus from the nutrients carried in the blood stream. After the nutrients for milk manufacture have been extracted from it, the blood is returned to the heart through the veins. This is a continuous process. The droplets of milk formed by the cells eventually pass via minute ducts into large ducts and hence to a cistern above each of the four teats.

The Health of Dairy Cows

Cows, just like human beings, need to be in a healthy condition to function at their best and the farmer takes great pains to ensure this. Cows are, however, subject to certain infectious diseases which, if uncontrolled, could have an adverse effect on the agricultural economy of the country. The Ministry of Agriculture, Fisheries and Food carries out clinical examination of dairy herds under various public health regulations.

Cows and Tuberculosis

Bovine animals are susceptible to tuberculosis and the organism can be excreted from a diseased animal either in the milk or in the coughed up sputum, thus constituting a risk to humans. For this reason a campaign to eradicate the disease from the national herd was carried out in the late 1940s and 1950s and was virtually completed in 1960. Now it is possible to declare that all milk sold in the United Kingdom comes from herds in which all the cows have been officially tested one by one and certified free from tuberculosis infection.

Tuberculin Test

The tuberculin test carried out in Great Britain is a comparative test using two tuberculins namely bovine and avian. The tuberculins are injected into each animal in a herd and the degree of reaction is measured 72 hours later. Animals which have given a positive reaction have to be removed for direct slaughter. Herds in which reactors are disclosed are subject to further tests at intervals of not less than 60 days until a clear test is obtained i.e. no reactors are disclosed.

The majority of herds in Great Britain are tested every three years but in certain parts of the country, mainly in the South West of England, the interval is two years or less.

Brucellosis

Brucellosis is another bovine disease which can be transmitted to man. One of its main effects in cattle is to cause a pregnant cow to abort. The loss of calves and drop in milk yield together with the disruption of the breeding programme can mean a serious financial loss to a farmer with the disease in his herd.

The first scheme to eradicate the disease, the Brucellosis (Accredited Herds) Scheme was introduced in 1967 and this was replaced in July 1970 by the Brucellosis Incentives Scheme. Both schemes were on a voluntary basis and provided a pool of accredited replacement (i.e. brucellosis free) stock to enable the Government to introduce compulsory eradication measures in 1972. The eradication programme has been most successful and by 1st November 1979 the whole of Great Britain will be subject to compulsory measures. It is expected that the disease will be virtually eradicated from the country by the early 1980s.

Mastitis

Mastitis is an inflammation of the udder caused chiefly by bacterial infection. Bacteria enter the udder through the teat duct and the disease is spread almost entirely at milking time from infected udders or teat sores. Infection is more likely if injury has occurred owing to the milking machine being in poor working order or to unskilled use by the operator.

Clinically affected cows have to be treated with antibiotics and the milk discarded for several days. Because of this and the reduced milk yield from infected cows, the disease costs the dairy farmer a lot of money (the farmer may not sell milk that is so contaminated).

Strict control of the disease is essential. A well trained milker using a properly serviced machine with a hygienic milking routine will ensure that the number of cases is kept to the minimum.

Hygiene on the Milking Premises

Milk drawn from the udder of a healthy cow contains very few bacteria. For this reason there is little risk of contamination if hygiene rules are rigorously applied.

The Buildings

Cows are milked in a milking house which may be

(a) a cowshed, i.e. a building in which cows may also be housed
(b) a parlour, which is used only for milking
(c) a milking bail, which is a movable structure used in the fields away from the farm building.

A milk room is a place in which the milk is cooled and stored; it is also usually used for washing and storing the milking equipment. For milk production it is essential for the buildings to be light, airy and well drained and have easily cleansed internal surfaces. The water supply must be adequate, as over 130 litres per cow per day may be used. On most farms the water comes from the public mains supply, but if water comes from a private source the source works are inspected and the water examined bacteriologically to ensure that it is satisfactory. Dung must be removed and the floor of the milking house cleansed every day. To minimise the risk of atmospheric contamination, matter containing dust, e.g. fodder and litter, must not be moved in the milking house during and immediately before milking.

The Cows

The cows must be clean. This is easier if long hairs on the flanks, tails and udders are clipped. The udders and teats are washed before milking using a hose and clean running water or a paper towel dipped in warm water to which a suitable disinfectant has been added. The udders should be dried after washing.

Equipment

Vessels containing milk must be properly covered. A supply of hot and cold water in the milk room is required. Equipment should be rinsed and washed with detergent and disinfected with boiling water, steam or an approved chemical agent.

The Farmer's Responsibility

The farmer's responsibility is to provide dairy workers with
(a) facilities for hand washing — hot and cold water
(b) an adequate supply of soap or suitable detergent, nail brushes and clean towels
(c) a suitable first-aid box containing bandages and antiseptics.

The Dairy Worker's Responsibility

The dairy worker's responsibility is to
(a) keep himself and his outer clothing clean and wear clean washable overalls and headgear
(b) keep any cuts or abrasions on exposed part of his skin covered and with a suitable waterproof dressing
(c) refrain from spitting or smoking.

Milking

Cows are usually milked twice daily — early morning and late afternoon. Their yields are highest soon after calving, often exceeding 20 litres per day, but yields of twice this amount are recorded from time to time. Yield remains high for a short period, then gradually falls off; the cow is usually 'dried off' about ten months after calving. The cow remains dry for approximately two months, during which time she is 'steamed up', i.e. given extra food in readiness for the next calving and lactation.

Milking Machines

Today milking machines are used on all but a very few dairy farms. They operate much as a suckling calf, vacuum from a vacuum pump is applied via the teat cup and to avoid pain and damage to the teat, vacuum and atmospheric pressure alternate usually once per second by means of a pulsation system.

The Milking Routine

The foremilk (the first milk to be drawn from the udder at each milking) is drawn off separately by hand from each cow into a strip cup. It is immediately visually examined in order that animals suffering from mastitis may be detected. The foremilk is rejected, the udder is washed, dried and a 'cup' is put on to each teat. The milk is drawn off into a closed bucket or pipe-lines to a refrigerated bulk vat.

Milking Parlours

There are several varieties of parlour, e.g. herringbone, tandem, but on a modern farm with a large dairy herd, you may find a rotary parlour some of which can be operated by one person. The cows stand on a rotating platform to be milked and as they leave the platform, others take their place.

Weighing and Recording

Many farmers weigh and record the milk from each cow. Official recording schemes are provided by the Milk Marketing Boards and regular samples are taken to measure butterfat and, in some cases, protein content. This helps efficient production because the farmer can feed according to the milk yield and so avoid 'under' or 'over' feeding. It also helps him to select the best animals for breeding.

Cooling

Milk leaves the cow at her body temperature and if left uncooled, it may turn sour very quickly. In the farm vat the milk is cooled and held at 4·5°C until it is collected.

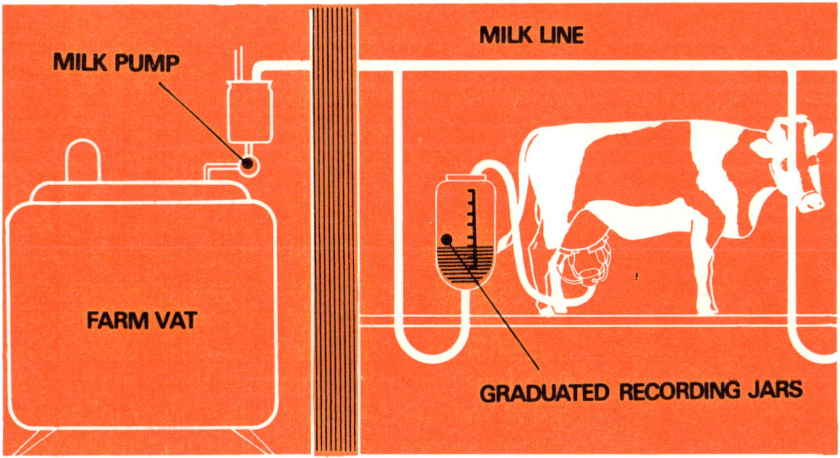

Collection

About 96 per cent of milk is now collected from bulk vats. A large insulated 'road tanker', which resembles a petrol tanker, calls for the milk which is checked for temperature (4·5°C), smell and appearance, mixed by an agitator, sampled and measured. The tanker has a special compartment at the rear carrying a polythene hose and suction pump. The hose is attached to the farm vat and the milk is sucked or pumped into the tanker. The milk from several farms is loaded into this vehicle.

Transfer to Depot or Creamery

The milk must arrive fresh so speed in transporting is essential and each area's milk takes the shortest route possible to its destination, which may be a country depot, a large town dairy or manufacturing creamery. Since a large proportion of our population is concentrated in conurbations remote from the milk producing areas, this milk can be directed to urban bottling plants in either a large bulk road tanker or rail tankers — this is referred to as reloading.

Milk is picked up in bulk from the farms in a tanker of 9,000 litre capacity, and is then transhipped into larger tankers of up to 20,000 litre capacity. This process enables larger quantities of milk to be moved greater distances using fewer vehicles. Where lesser distances are involved, the farm collection tankers do not reload but deliver direct to the processing dairy.

The milk from several farms may be mixed and it would not normally be possible to return the milk to the farmer. In these cases milk which does not pass the necessary laboratory tests on arrival at the dairy will be rejected and the financial loss borne by the Milk Marketing Board. (See page 22).

Tanker Cleaning

Various automated methods of cleaning tankers are used by the dairy. A typical process is:

1. rinse with mains water
2. circulate warm detergent (or detergent steriliser) solution
3. rinse with water
4. circulate an approved sterilising solution
5. rinse with clean water.

Milk Marketing Boards

All milk sold off farms in the United Kingdom must be sold through the appropriate Milk Marketing Scheme.

There are five Milk Marketing Boards in the United Kingdom, one for England and Wales (set up in 1933), three for Scotland (set up in 1933) and one for Northern Ireland (set up in 1955). Established under the Agricultural Marketing Acts, the Schemes administered by the Boards must be approved by Parliament, but the MMBs are producer-controlled and Board members are elected representatives of milk producers together with representatives appointed by the Minister of Agriculture, Fisheries and Food.

The Milk Marketing Boards arrange contracts for the sale and collection of milk from the farm and payment to the farmer. About 96 per cent of the milk is sold wholesale either to the Board (in England and Wales, and in Northern Ireland), or (in Scotland) to a buyer under an arrangement to which the Board is a third party. The remainder is sold, under licences issued by the Boards, by producer-retailers direct to consumers or distributors.

The prices at which the Boards or producers sell milk for liquid consumption are subject to Government approval and in practice are determined by the Government. Milk surplus to the requirements of the liquid market is sold by the Boards for manufacture into dairy products at lower prices negotiated by the Boards with the manufacturers. Accordingly, the prices producers as a group receive for the milk depend on the pooled realizations from the different markets within the framework of the guaranteed price structure, but the price received by an individual producer also depends on the compositional and hygienic quality of his milk, and is independent of the market in which it is sold. The maximum prices at which liquid milk may be sold are prescribed by the Government.

The Boards are responsible for organizing the flow of milk from farms to town dairies, country depots and creameries; for the collection of milk from farms either by the Boards' transport fleets or by hauliers operating under contract; and for administering the system of Government-fixed margins and allowances relating to liquid milk movements and processing.

In addition, the Boards offer a variety of other services to milk producers such as the Artificial Insemination, Milk Recording, and advisory services to improve cattle breeding and to help farmers to become more efficient.

Chapter III

MILK PROCESSING AND DISTRIBUTION

Heat Treatment of Milk

Ninety-six per cent of milk sold in the United Kingdom today by retail is heat treated — that is pasteurized, ultra-heat-treated or sterilized.

All apparatus and methods used for heat treating milk must be approved by the licensing authority.

Pasteurization

Pasteurization kills any pathogenic organisms which may be present, without significantly affecting the nutritional value of the milk. It is carried out either by heating milk to 63°C—66°C for half an hour, or (more usually) by heating to not less than 72°C for at least 15 seconds. As soon as heating ends, the milk is cooled rapidly to not more than 10°C. The apparatus used for this is called a *heat exchanger*. These time/temperature combinations are laid down in the Milk (Special Designation) Regulations 1977.

In the heat exchanger system (see the diagram on page 65) there is a balance tank to which raw milk flows from the storage tanks and in which a constant level of milk is maintained by a special float valve. From the balance tank the milk is pumped at a rate regulated by a flow control valve to the regenerative section of the heat exchanger. Here it is pre-heated by milk which has previously been pasteurized and is now being returned for cooling. The pre-heated milk passes to the filter and then to the heating section where it is heated to a temperature, just above the legal minimum for the pasteurizing process, by the circulation of hot water only 1°C—2°C above the temperature of the milk.

The immediate rapid cooling of the milk within the enclosed system is an essential part of its treatment since to keep it at a high, or warm, temperature would encourage the growth of heat-resistant organisms and tend to decrease the nutritional value (see page 32). In addition, when milk has not been homogenized, delay in cooling adversely affects the cream line. The cooling is done partly by regeneration (the process described above which also pre-heats the incoming raw milk), partly by cold water and finally by chilled water or brine.

If the pasteurizing temperature in the exchanger drops below the minimum a special valve comes into operation which prevents the milk going forward and returns it to the balance tank for re-processing.

Although most milk is filtered to remove sediment and extraneous matter, there is an increasing use of clarifiers in which milk is spun at very high speeds and the sediment removed by centrifugal force.

Packaging

After pasteurization the milk must be bottled or packed in containers as soon as possible. Until then it is stored in large insulated stainless steel tanks.

Bottle Washing

Before passing to the bottle-filling machines, the bottles are washed by machine. The bottles are given a pre-rinse with tepid water. This is followed by washing in hot detergent solution, followed by a series of warm and cold water rinses.

Bottles may be washed by a 'soaker' system whereby they pass slowly on a conveyor through a solution tank and are completely submerged and subsequently rinsed with clean water. The alternative is the 'hydro' system by which the treatment liquid is jetted violently into the bottles from a tank below the conveyor line and returned to the same tank. A combination of both systems is frequently used. The bacteriological condition of washed bottles is subject to strict control and bottles leaving the machine are tested for sterility at regular intervals.

Bottling

The bottling is usually done by a high-speed machine which takes a steady stream of thoroughly cleaned bottles from a moving belt, automatically fills them with the correct amount of milk and seals them with metal foil caps. Modern bottling machines can fill and cap 600 bottles a minute. All bottle tops are now required by law to fit tightly over the lip of the bottle and there are also regulations governing the colour of aluminium foil which can be used to cap various types of milk (see page 21).

Bottles

Milk bottles are the property of the dairy supplying them and should be returned promptly in a reasonable condition. The average life of a milk bottle is about 24 deliveries.

Non-returnable Containers

Increasing use is being made of non-returnable containers for milk, but because they are used only once the cost is higher than for bottles. For this reason the dairy may charge more for milk sold in this way. Added to this, the actual filling is a slower process and thus involves greater labour costs.

A number of such containers are on the market including paper and plastic cartons, bottles and sachets. There is a wide range of shapes and sizes for domestic and large-scale use.

Other forms of Heat Treatment

Ultra Heat Treatment

The ultra heat treatment of milk (UHT) is carried out in an apparatus similar to that used for pasteurization but the milk is heated to at least 132°C for 1 second. When packed under sterile conditions into suitable containers it will keep for several months even without refrigeration (as long as the container is unopened) with minimum effect on the flavour.

Sterilization

Sterilization destroys bacteria and other micro-organisms more completely than pasteurization.

There are no exact methods laid down for the procedure to be followed but the milk is first pre-heated, homogenized (see page 20) and filled into glass bottles which are closed with a hermetic seal. The final process of 'sterilization' may then be carried out as a batch or as a continuous process.

In the *batch process* the filled bottles are heated for 15 to 40 minutes to a temperature of 104°C—113°C in autoclaves. They are then removed and left to cool naturally in the atmosphere.

In the *continuous process* the bottles pass on a conveyor belt through hot water tanks, via a water seal into a pressure steam chamber, for 15 to 40 minutes at 107°C—113°C, and out via a second water seal into a series of water-cooling tanks.

The higher temperature used in sterilization causes slight caramelization of the lactose producing a 'cooked' flavour and more creamy appearance. Sterilized milk will keep for two or three months without refrigeration as long as the container is unopened.

Homogenization

Strictly speaking, the term homogenized means 'made uniform' and milk is homogenized to give a product which remains of uniform composition (the cream does not rise to the top).

In ordinary milk the fat globules vary in diameter from about 1μ (1/1000 mm) to about 18μ—the average being 4μ. In the process of homogenization the milk is forced through a tiny orifice under considerable pressure. This breaks up the fat globules to a uniform size of from about 2μ to about 1μ in diameter —according to the pressure of homogenization.

Homogenized milk has a smoother, creamier taste and it is more readily digested — partly because of the smaller fat globules and partly because the process lowers the curd tension which results in the formation of a softer curd during digestion.

Milk Distribution

Milk may go from the bottling dairy direct to consumers or to sub-branches of the same firm or to independent retailers. There are over 40,000 milk roundsmen in the United Kingdom. In addition to the distributors there are just under 4,500 producer-retailers—farmers who produce milk for direct sale to their own customers. Most liquid milk is sold in the towns and cities— some 18,000,000 households, as well as hospitals, factory canteens, hotels and restaurants.

Types of Milk

There are four statutory designations covering all milk. Their use is compulsory by law and the grade is indicated by the printed wording on the cap or container.

The designations are:

Untreated Milk—where the raw milk is not subjected to any form of heat treatment. It may be produced and bottled by a farmer holding a licence from the Ministry of Agriculture, Fisheries and Food, or bottled by a dealer holding an 'untreated' licence from the local authority. This milk must come from Brucellosis Accredited herds.

Pasteurized Milk — milk which has been subjected to the pasteurization method of heat treatment (see page 18).

UHT Milk — milk which has been subjected to ultra heat treatment (see page 19).

Sterilized Milk — milk which has been heat treated to above boiling point for such a time as to ensure that it will comply with the Turbidity Test (see pages 19 and 23).

In addition there are other kinds of milk found within these four special designations. These are as follows:

Homogenized Milk — milk which has been heat treated and so processed as to break up the tiny globules of milk fat and distribute them evenly throughout the milk instead of allowing them to rise to the top. The process prevents the formation of a 'cream line' at the top of the bottle (see page 20).

Channel Islands Milk — milk from cows of the Channel Islands breeds (Jerseys and Guernseys) with a minimum butterfat content of 4 per cent.

South Devon Milk — milk from cows of the South Devon breed with a minimum butterfat content of 4 per cent.

Colour Code for Bottle Caps and Lettering

The Milk and Dairies (Milk Bottle Caps) (Colour) Regulations 1976, which apply to England and Wales, prescribe against specified descriptions of milk, the colours to be used for aluminium foil caps and for any non-embossed lettering on them. The types of bottled milk and statutory colours include:

Type of Milk	*Colour of Cap*	*Colour of Lettering*
Pasteurized Channel Islands and South Devon	Gold	Black or Silver
Pasteurized Homogenized	Red	Black or Silver
Pasteurized	Silver	Black
Untreated Channel Islands and South Devon	Green with Single Gold Stripe	Black or Silver
Untreated	Green	Black or Silver
Sterilized	Blue	Black or Silver
Ultra Heat Treated	Pink	Black or Silver

Chapter IV

QUALITY CONTROL OF MILK

The importance of the quality control of milk at all stages from production through transport, reception, processing and distribution has long been recognised and, in fact, the dairy industry was foremost in the food industry in establishing control laboratories staffed by trained personnel.

Quality assessment became more co-ordinated from 1942 following the introduction of the National Milk Testing and Advisory Scheme. A wide variety of tests are now carried out to ensure that the raw milk is of a suitable standard and that heat treated milk meets all the stringent standards for marketing a high quality product.

In addition to the tests carried out on milk, a wide range of tests are applied to ensure that all equipment and containers are satisfactory and cleaning and processing procedures are correctly carried out.

Principal Tests

Hygiene Test

All milk supplied by producers to dairies is checked at regular intervals to ensure that it is produced under hygienic conditions and therefore does not contain too many bacteria. Samples from bulk farm tanks are examined four times a month. A dye reduction test is used for this purpose. The sample of milk is stored in a refrigerator overnight and then 1 ml of the dye resazurin is added to 10 ml of milk which is then incubated in a water bath at $37 \cdot 5°C$ for two hours. The dye is affected by bacteria which cause its original violet colour to change to pink and finally to colourless. The rate at which this change occurs is dependent upon the number and type of bacteria present. The colour of the milk plus dye is compared after incubation for the appropriate period with a known standard using a comparator which contains a disc carrying numbered colour shades. The corresponding shade and therefore the relevant number gives some indication of the bacterial condition of the milk. If the milk falls below a certain standard, it fails the test and a price reduction may be imposed on the producer. There is an 'early warning' system under which a producer will be notified if his supplies are approaching failure level. This test is applied in England and Wales. Scotland and Northern Ireland use a modified hygiene test.

When supplies arriving in bulk tankers at a dairy are suspected of being sour or produced under very unhygienic conditions, they may be subjected to the *rejection test* which requires the milk to be incubated with resazurin for 10 minutes and those giving a disc reading below a certain level will be rejected by the dairy.

Gerber Test

The Gerber test is used to determine the *fat content* of milk, and a Gerber butyrometer — a small glass tube with a bulbous portion ending in a calibrated scale — is required.

First 10 ml of sulphuric acid of density: 1·815 g/ml at 25·5°C are placed in the tube, then 10·94 ml of a sample of milk (thoroughly mixed) are added by means of a pipette, and finally 1 ml of amyl alcohol is added. A stopper is placed firmly in the neck of the tube. The tube is shaken and the milk curd is dissolved by the action of the acid.

The principle of this is that the amyl alcohol will help to release the fat and the acid will dissolve everything but the fat which will then rise into the now up-ended and calibrated part of the tube. Rather than wait for this to happen under normal force of gravity the Gerber tubes are centrifuged for not less than four minutes at 1,100 r.p.m. On removal from the centrifuge the tube is warmed in a water bath at 65°C and the fat fraction can then be read as a direct percentage. Many dairies now use automatic testers in which results can be obtained in a few seconds.

Hydrometer Test

For normal control purposes the density of the milk sample is obtained by means of a hydrometer. This density is then related to the known fat content and, by means of a formula, the total *solids-not-fat* can be calculated.

Phosphatase Test

This test is carried out to check *whether the milk has been adequately pasteurized* and depends on the fact that an enzyme called phosphatase is inactivated by effective pasteurization.

1 ml of a sample of milk is added to 5 ml of a solution of a complex chemical, disodium *p*-nitrophenyl phosphate. Phosphatase, in its active form, has the power to break down this chemical into two separate substances. One of these is *p*-nitrophenol which is yellow and which, if present, will colour the milk. The degree of colouration is compared with known standards using a comparator.

Methylene Blue Test

The methylene blue test is used to assess the *keeping quality of the pasteurized milk*. Samples are first stored overnight, either at atmospheric shade temperature during the summer months or at 18·3°C during the winter. Then 1 ml of a dye, methylene blue, is added to 10 ml of milk. This reagent works in the same way as resazurin but the colour change goes from blue to colourless.

Samples are put in a water bath at 37·5°C for half an hour, and provided that the milk has been processed, bottled and distributed under hygienic conditions there will be no colour change at the end of this time.

Turbidity Test

The turbidity test is used to check the *efficiency of sterilization* and depends on the fact that, when milk has been properly sterilized, the protein structure changes. As a result of this change, the protein is precipitated when ammonium sulphate is added to the milk sample.

The milk is then filtered and a completely clear solution should result when the filtrate is heated. Any failure in sterilization will leave some of the protein unchanged and a turbid effect will be seen in the heated filtrate.

Colony Count Test

This test is carried out to check that *UHT milk has been properly heat treated* and therefore contains no or very few bacteria. A small amount of the milk is mixed with sterile yeastrel milk agar (a nutrient medium which supports bacterial growth) and incubated between $30°C$ for 72 hours or $37°C$ for 48 hours.

After this time the purity of the milk can be assessed by counting bacterial colonies present.

Antibiotic Tests

Two principal bacteriological tests are used for the detection of antibiotics. In England and Wales and in Northern Ireland the test used is a version of the BCP test. This test depends on the fact that in the presence of a dye BCP (bromocresol purple) the growth of a sensitive organism, Streptococcus thermophilus, results in a change of colour from blue through green to yellow. In the presence of antibiotics in sufficient quantity, the growth of the organism is inhibited and there is no colour change. The sample of milk is first heat treated to destroy any natural inhibitory substances and is then incubated in the presence of the organism, nutrients and bromocresol purple indicator at a temperature of $37·5°C$ for four hours. The presence of antibiotic or other inhibitory substance is indicated if the incubated sample remains blue.

In Scotland a disc assay method is used for the detection of antibiotics and other inhibitory substances. In this test, a small disc of filter paper is dipped into the milk sample and is placed on the surface of an agar medium contained in a Petri plate and inoculated with a sensitive organism Bacillus calidolactis. The plate is incubated at $55°C$ for $2\frac{1}{2}$ hours during which time growth of the test organisms causes the agar medium to become cloudy. Antibiotics or other inhibitory substances, if present in the milk sample, pass into the agar medium in the vicinity of the disc, preventing the growth of the test organism, and this results in the retention of a circular clear zone around the disc. The size of the clear zone is related to the type and concentration of the antibiotic or other inhibitory substance present in the milk.

Sediment Test

Where milk is produced under unhygienic conditions, the quantity of sediment present will increase. An assessment of the amount of sediment is made by filtering a fixed amount of milk under pressure through a cotton pad and comparing this against known standards. Appropriate action can then be taken to improve supplies where necessary.

Freezing Point Test

The presence of extraneous water in milk must be avoided at all costs. The freezing point of genuine milk falls in a narrow range: $-0·530°C$ to $-0·550°C$ with an average of around $-0·544°C$. The presence of extraneous water results in the freezing point rising nearer to $0°C$ and since this change is proportional to the amount of water present, it is possible, by determining the freezing point of the milk, to calculate the quantity of extraneous water, if any. Modern instruments allow this determination to be made to the nearest ·001°C on a small quantity of milk in two or three minutes.

Milk Quality Control Schemes

The Joint Milk Quality Committee consists of representatives of the Milk Marketing Board, the Dairy Trade Federation,* the National Farmers' Union and observers from the Ministry of Agriculture, Fisheries and Food, and of the Department of Health and Social Security. This organization administers schemes to control the marketability of milk supplies.

Milk of poor *bacteriological quality* is subject to price deductions for the producers if it fails the hygiene test. Producers who repeatedly supply poor quality milk are reported to the Milk Marketing Board who may cancel their contract.

Tests for *compositional quality* — the total solids content of the milk — are also carried out at regular intervals. An annual average of the milk supplied is calculated and a scale of over 40 classes allows for price differentials to be paid to producers according to the compositional quality of their milk. Those supplies with an annual average of 8·5 per cent or more solids-not-fat and 12·4 per cent to 12·5 per cent total solids receive the basic price. Those above receive graduated higher payments and those below smaller payments. In addition those with average solids-not-fat below 8·5 per cent are subject to further price deductions.

A price deduction scheme is also operational for milk found to contain *antibiotics or other inhibitory substances* (including dairy sterilizing agents, disinfectants, germicidal ointments, etc.). Consignments of milk from farms are tested at regular intervals by the dairy receiving them and failures are reported to both the producer concerned and the Milk Marketing Board. Price deductions may be imposed on consignments of milk that fail the test.

The Dairy Trade Federation is the negotiating body that deals with matters relating to the purchase of milk from producers and subsequent manufacture and distribution. It represents four major associations — the Amalgamated Master Dairymen, the Co-operative Milk Trade Association, the National Association of Creamery Proprietors and Wholesale Dairymen and the National Dairymen's Association.

Chapter V

LEGAL CONTROL

Legislation of milk is strict and comprehensive. It covers every aspect from the health of the cows on the farm to the capping of the bottles at the bottling plant and the condition in which the product reaches the consumer.

Landmarks in the Legal Control of Milk in England and Wales

1. Public Health Act 1875.
2. Sale of Food and Drugs Act 1875.
3. Food and Drugs Act 1899.
4. Sale of Milk Regulations 1901.
5. Milk (Special Designations) Order 1922.
6. Tuberculosis Order 1925.
7. Milk and Dairies Order 1926.
8. Attested Herds Scheme 1934.
9. Milk (Special Designation) Order 1936.
10. Sale of Milk Regulations 1939.
11. Milk (Special Designation) (Pasteurized and Sterilized Milk) Regulations 1949.
12. Milk (Special Designation) Raw Milk Regulations 1949.
13. Tuberculosis (Attested Herds) Scheme 1950.
14. Food and Drugs Act 1955.

Government Regulations now in force in England and Wales*

1. *The Sale of Milk Regulations 1939,* which prescribed the presumptive minimum butterfat content of milk as 3 per cent and presumptive minimum solids-not-fat content as 8·5 per cent lapsed on 1st January 1976. These have been replaced by the *Drinking Milk Regulations 1976,* to be read in conjunction with Council Regulation (EEC) No. 1411/71 as amended by Council Regulation (EEC) No. 566/77, which apply only to milk sold for human consumption and provide that the following types of drinking milk may be delivered to consumers:

a. *Raw (untreated) milk* with its composition unaltered.

b. *Non-standardized whole milk produced in the United Kingdom*, with its composition unaltered since the milking stage but subject to a minimum fat content of 3 per cent. Although Channel Islands and South Devon milk are included in this type, the law regarding these milks is unchanged and it is an offence to sell them with a fat content of less than 4 per cent, under the Milk and Dairies (Channel Islands and South Devon Milk) Regulations 1956.

Regulations for other areas are listed in Appendix 1.

c. *Standardized whole milk* imported from other Member States of the European Economic Community with a fat content of not less than a guideline figure, fixed by the Council of the European Communities, equivalent to the weighted average fat content of whole milk produced in the United Kingdom during the previous year. This is currently 3·77 per cent but may vary slightly each year.

d. *Semi-skimmed milk* with a fat content of at least 1·5 per cent and not more than 1·8 per cent.

e. *Skimmed milk* with a fat content of not more than 0·3 per cent.

   Types b, c, d and e are all required to have been heat treated.

2. *The Milk and Dairies (General) Regulations 1959* deal with the registration of dairy farms, farmers and distributors and lays down strict standards of hygiene for buildings and employees which must be adhered to; the Ministry of Agriculture, Fisheries and Food is responsible for enforcement of the Regulations on the farm and the local authorities in the dairies.

3. *The Tuberculosis Order 1964* lays down the requirements necessary for a herd to be registered as Attested. (All herds are now attested i.e. certified tuberculosis free.)

4. *The Milk (Special Designation) Regulations 1977* provide for licences for the production and sale of specially designated milks and prescribe the conditions under which licences are granted. One of these conditions is that the milk which carries the special designation "Untreated" must have been produced from a Brucellosis Accredited Herd.

5. *The Food and Drugs Act 1955* makes it an offence to "sell food at the prejudice of the consumer which is not of the nature or not of the quality or not of the substance demanded". It also makes it an offence to add water (and certain other things) to milk intended for sale for human consumption.

6. *The Milk and Dairies (Milk Bottle Caps) (Colour) Regulations 1976* prescribe against specified descriptions of milk the colours to be used for aluminium foil caps and for any lettering on them.

Enforcement of Regulations

Regulations are enforced either by the Ministry of Agriculture, Fisheries and Food or by local authorities.

On Farms

All herds are tested regularly for tuberculosis and the tests are enforced by law. Dairy Husbandry Advisory Officers of the Agricultural Development and Advisory Service (part of the Ministry of Agriculture, Fisheries and Food) visit farms to advise on milk production and handling and to see that the buildings, water supply and methods comply with the *Milk and Dairies (General) Regulations 1959.* Officers of the Agricultural Development and Advisory Service also advise on correct food, herd management, etc.

No hard-and-fast-rules are laid down concerning the number of times a milk producer will be visited in a given time but on average this works out at once in three years. When unsatisfactory conditions are found further follow-up visits are always made. If conditions continue to be unsatisfactory, the farmer may lose the right to produce milk or he may be fined.

More frequent visits are made to producers holding licenses to sell untreated milk and, in these cases, milk samples are taken at least four times a year. In the event of a sample failing tests, further samples are taken.

Between the Farm Gate and the Consumer

Once the milk has left the farm the public health responsibility for enforcing the law is taken over by the local authorities who work closely with the dairies and manufacturing creameries. The enforcement of regulations dealing with the *composition and labelling of foodstuffs* is also the responsibility of the local authorities.

Environmental Health Inspectors carry out the day-to-day work of the Health Department under the Food and Drugs Act. In addition, in some provincial towns, the local Boards of Health and Town Commissioners have special powers. A milk and dairies inspector must have a complete knowledge of dairy work, its hours of work and methods of business, as well as experience of the technical details of the trade, such as the operation of heat treatment plants. Equipment may be checked at any time and samples of food may be taken at any stage. These include all points of sale to the consumer such as shops, milk delivery vehicles, vending machines, cafés and restaurants. Environmental Health Inspectors and other local authority officers take samples of milk to check pasteurizing and sterilizing treatments and they take routine samples for determining the presence of antibiotics as well as brucellosis and bovine tuberculosis.

Precise methods of sampling are laid down* to meet legal requirements and the sampling officer must inform the person from whom a sample of food has been taken of his intention to have a sample analysed. He must then and there divide the sample into three parts. One is handed to the seller, one is passed to the analyst and the third kept for future reference should a dispute arise as a result of court proceedings.

Food and Drugs Act 1955, Milk (Special Designation) Regulations 1977.

Chapter VI

COMPOSITION AND NUTRITIONAL VALUE OF MILK

Milk is the most nearly perfect food containing as it does some of all the nutrients necessary to maintain life and promote growth — proteins, fats, carbohydrates, minerals and vitamins.

Proteins are essential for growth and repair of the human body; *fats and carbohydrates* supply the body with energy and warmth; *minerals* are vital to the building of bone structures and teeth and the regulation of the body processes; *vitamins* are needed to maintain good general health. All these are present in milk in a form which is readily digested and ready for immediate use by the body — there is no waste.

Composition of Milk

Milk is an emulsion — a suspension of minute droplets of fat in a watery solution which contains proteins, lactose (milk sugar) and mineral salts. The milk of several species of animals is used for human consumption in different areas of the world, but this handbook deals only with cows' milk.

The colour of milk may vary from bluish-white to golden yellow, depending on its composition and the amount of yellow pigment (carotene) which is present. These in turn depend on many factors including the breed of cow, the food she consumes, the period of lactation, her individuality and state of health, etc.

When the milk from a large number of cows is bulked together, the individual variations cancel out so that there is relatively little difference in nutritional value between one large consignment of milk and another.

Nutrients in Milk

The greater part of milk — just over 87 per cent by weight — is water. The remainder is made up of solids which may be divided into two main groups, the milk fat (butterfat) and the solids-not-fat (S.N.F.).

The following table shows the average composition of one pint of pasteurized milk sold in the United Kingdom.

Protein	19·3g		Vitamin B Group:	
Fat	22·2g		Thiamin	0·23mg
Carbohydrate	27·5g		Riboflavin	1·11mg†
Calcium	702mg		Nicotinic Acid equivalents	5·03mg
Iron	0·3mg (negligible amount)		Vitamin B_{12}	1·8μg
Phosphorus	540mg		Pantothenic Acid	2mg
Kilocalories (Kilojoules)	380 (1592)		Biotin	11·4μg
Vitamin A	220μg (Summer)		Vitamin C	6mg‡
(retinol equivalents)	160μg (Winter)		Vitamin D (Summer)	0·17μg
			(Winter)	0·07μg

† *Falls on exposure to sunlight and on heating*
‡ *Falls on storage: content for pasteurized milk after 12 hours storage*

The butterfat portion is of great commercial importance because of its taste and its use in the manufacture of butter and cream. The butterfat content is often used by the public to assess the value of milk supplied, but nutritionally the solids-not-fat portion is more important because it contains the proteins, lactose, mineral salts and water-soluble vitamins.

Butterfat

The fat in milk is suspended in the 'milk serum' in the form of minute globules. They vary in diameter, the average size being 4μ. Milk fat, being lighter than water, rises to the top of the milk as cream. The larger the fat globules the more rapidly and more completely the cream rises to the top and churns more easily in buttermaking. Reducing the size of the globules is the principle behind homogenization which results in the even distribution of fat throughout the milk.

The size of the fat globules varies with different breeds of cow. Channel Islands breeds produce milk with larger fat globules than other breeds such as Friesians, Shorthorns or Ayrshires.

Certain chemical changes in butterfat are responsible for taints in milk, cream and butter. These changes are sometimes caused by fat-splitting enzymes (lipases) which may free fatty acids and cause rancidity. Butyric acid will produce very rancid flavours in butter.

Proteins

The proteins in milk are of a very high biological value because they contain all the essential amino-acids required by the human body. The principal milk proteins are casein, lactalbumin and lactoglobulin. *Casein* is found only in milk and milk products. In milk it is present in colloidal form and is precipitated at a pH of 4·6. This is a process known as coagulation and occurs in the presence of acids, as when milk enters the stomach or when it is allowed to go sour. A similar precipitation can occur without the presence of acid, when rennet is added; this is usually called clotting rather than coagulation. *Lactalbumin* and *lactoglobulin* are similar to albumin and globulin found in the blood. Unlike casein, they are not precipitated during cheesemaking but are left in the whey.

Lactose

Lactose, or milk sugar, a disaccharide composed of glucose and galactose combined is the only carbohydrate present and is found only in milk. It is less sweet in taste than sucrose (cane or beet sugar) and is readily fermented by the action of lactic acid bacteria, thus producing lactic acid. This acid is the primary cause of milk turning sour. At 0·3—0·4 per cent lactic acid there is a sour taste; at 0·6—0·7 per cent lactic acid the milk curdles.

Minerals

Milk contains some of almost all the mineral elements and salts which the human body requires, and it is extremely rich in calcium. The other main mineral constituents are phosphorous, potassium, sodium and chloride; traces of iron, magnesium and sulphate are also present, but the amount of iron is insufficient for human requirements.

Just under one third of the minerals in milk are in true solution, the remainder are associated with the milk solids. From a nutritional point of view the most important of these is calcium; milk and cheese are two of the most important sources of this nutrient in human diets.

Vitamins

Milk is particularly important as a source of riboflavin in the diet, and contains appreciable quantities of vitamin A and thiamin; it is comparatively deficient in vitamins C and D. Vitamins A and D are fat-soluble and are found in the milk fat, while those of the B group and vitamin C are water-soluble and thus found in the water and solids-not-fat portion.

Exposure of milk to daylight and sunlight results in losses of riboflavin and vitamin C. The latter is colourless and does not absorb visible light, but riboflavin, which is yellow, absorbs blue light, and this acts as a sensitizer bringing about the photochemical oxidation of vitamin C. It is therefore important not to leave milk on the doorstep for too long.

Micro-organisms in Milk

Raw (untreated) milk will undergo a natural souring followed by a separation into a solid, curd, and a liquid, whey. This change is caused by micro-organisms which gain access to the milk. The chief of these is the *lactic acid bacteria* because the action of this micro-organism converts lactose into lactic acid. The curd consists of the bulk of the proteins and the fat; the whey consists of the lactose, a small amount of the proteins, minerals and water. In process of time the protein decomposes, owing to the access of putrefactive micro-organisms.

Care should be taken not to confuse the natural souring micro-organisms with either the putrefactive or pathogenic ones. The action of the first is utilized in the making of cheese and other dairy products while the possible dangers of the other two are virtually eliminated by strict hygienic control and the heat treatment of milk.

Variations in Requirements of Nutrients

Requirements of the nutrients vary with age, sex, type of work, a person's individuality, etc. Milk is a food which, more than any other single food, can supply most of these requirements.

Expectant and nursing mothers require a diet which is rich in protein and body-building foods to form strong bones and teeth, firm muscles and healthy tissues. Milk is a particularly important food at this time as it is a very good source of calcium and protein. They need at least one pint of milk a day.

Infants from birth to one year should have their diets supervised by medical advisers. The food given in their first few months will almost certainly contain a high proportion of milk — either human milk or modified milk from cows. Notes on various types of artificial foods for infants will be found on pages 34 to 35.

Children from birth to five years are passing through probably the most important years of their lives as it is then that the foundations of future health are well and truly laid. Milk is of special importance to this group which requires large amounts of calcium and protein for body-building and growth.

Young schoolchildren from five to eleven years are going through a period of intense mental and physical activity as well as growth. Omission of milk from the diet could lead to soft bones, poor teeth and an increased liability to illness.

Adolescents from eleven to fifteen years are going through a period of accelerated growth as well as physiological and psychological changes. Their need for calcium at this time is particularly great and milk is therefore of major importance.

Adults continue to obtain an important proportion of their dietary intake of protein, calcium, riboflavin and, to a lesser extent, vitamin A from milk.

For elderly people, often too tired to prepare a proper meal for themselves, milk is very important because a glass is almost a meal in itself. A pint of milk supplies the daily calcium requirements for adults.

Contribution of Milk to the Diet

Foods which supply one-sixth ($16\frac{2}{3}$ per cent) of one's daily requirements of any nutrient are considered to be a reasonable source of that nutrient. The following table shows the proportion of daily requirements supplied by a pint of milk.

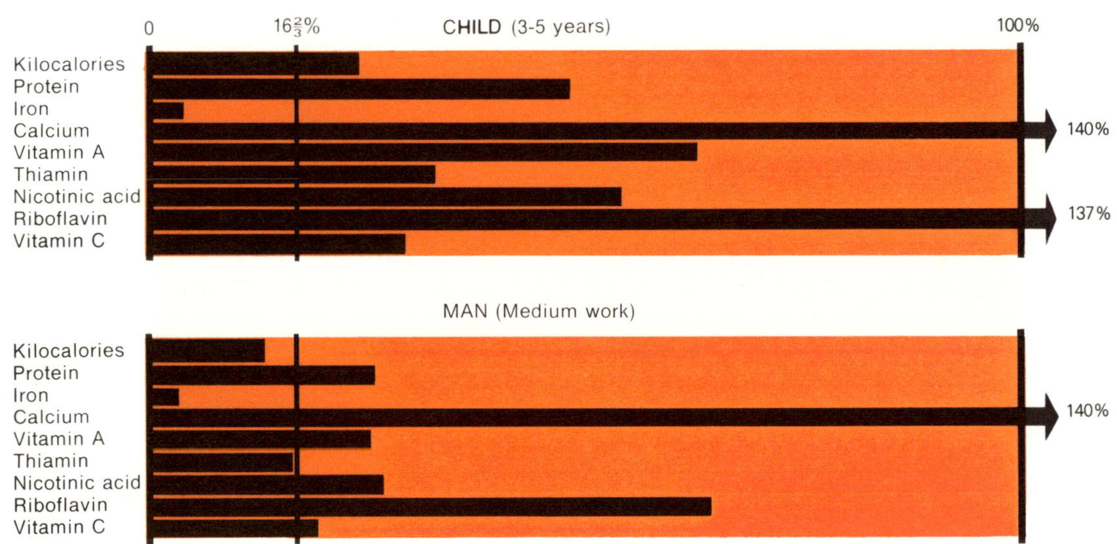

Based on a table in "Recommended Intakes of Nutrients for the U.K." (H.M.S.O. 1969).

Effects of Heat Treatment on Nutritional Value

Pasteurization is a mild form of heat treatment which destroys disease-causing bacteria and causes a slight decline in nutritive value. Up to 10 per cent of the thiamin and vitamin B_{12} are destroyed and 25 per cent of the vitamin C. The effects of *drying* need not be much greater than those of pasteurization and nutritionally the best roller-dried powders are almost equal to high-quality spray-dried products (see page 52). The effect of *ultra heat treatment* on nutritional value is similar to that of pasteurization.

The more severe heat treatment required in *sterilization* causes a loss of about 60 per cent of the vitamin C, 20 per cent of the thiamin and nearly all the vitamin B_{12}. There is also a slight reduction in the biological value of the protein.

Preparation of evaporated milk results in a loss of about 60 per cent of the vitamin C and 40 per cent of the thiamin. Preparation of *sweetened condensed milk* involves lower temperatures and the loss of nutritive value is smaller — similar to that which occurs in pasteurization.

Economic Value of Milk

For practical purposes, the nutritional value of food must be considered in relation to its availability and price. Milk is important in these categories too, for it is readily available at all times in this country and it is one of the most economical of all good quality foods. There is no waste — every drop can be used.

Comparisons with other foods (raw) emphasizes this point. For example, the protein in one pint of milk is equivalent to that in three medium eggs (grade 4), or just over 90g of liver and stewing steak or 110g of white fish; the calcium in one pint of milk is equivalent to that in nearly 90g of cheese, 23 medium eggs or 17·5kg of potatoes; the energy value is equivalent to that in 500g of white fish, four medium eggs or nearly 220g of stewing steak. (Calculations are based on tables in *'Manual of Nutrition'*, 8th edition revised, 1978).

Chapter VII

MILK IN THE HOME

The United Kingdom is one of the very few countries in the world where milk is delivered regularly to the doorstep. Great care is taken to ensure that it is in good condition and unadulterated on arrival and similar care should be taken in the home.

Care of Milk

It is important to take milk indoors as soon as the milkman delivers it; if this is not possible, arrangements should be made for it to be put in a cool place, where birds are unable to peck the metal caps and other damage can be prevented. Milk should never be left in direct sunlight, as this affects its flavour and destroys part of its vitamin content.

When there is no refrigerator, milk should be kept in the coolest place available — if possible in a current of cold air. The floor is cooler than a table or upper shelf. It should never be put in an unventilated cupboard.

A good way to keep milk cool is to put the bottle in a basin half filled with cold water and to place over it a clean flower-pot or a piece of butter muslin saturated with cold water, making sure that this reaches the water-level. The water should be changed regularly. Milk must be kept in a scrupulously clean container. It will keep best in the bottle in which it is delivered for this has been sterilized before being filled. Jugs which have contained milk should first be rinsed in cold water and afterwards washed and scalded with boiling water. They should be allowed to cool and drain before being used again. Milk jugs should be covered so that dust and flies cannot carry harmful bacteria into the milk. Even in a refrigerator, it is advisable to cover milk to prevent it absorbing flavour from other foods. Yesterday's milk should never be mixed with today's.

Bottles

Milk bottles should be rinsed in cold water as soon as they are empty and put outside promptly, in a safe place, for collection. The bottles are the property of the dairy supplying them and should not be used for anything but milk. Furthermore, if paint, turpentine, hair lacquers, permanent wave lotions or similar products, are put into them no bottle-washer can remove the residue; they become as unacceptable as if chipped or broken.

Artificial Feeding of Infants

It must be stressed that the feeding of very young infants should be governed by advice from the doctor, midwife, health visitor or nurse who is in charge. The following notes are merely a general guide.

The composition of cows' milk is different from breast milk. For this reason it is now recommended that unmodified cows' milk or preserved whole milk products such as evaporated milk and dried full cream milk, is not used for feeding infants less than nine months old.

Composition of Human and Cows' Milk

Nutrient	Grams per 100g		
	Human Milk (10 days post partum)	Human Milk (Mature)	Cows' Milk
Protein	2·0	1·3	3·3
Fat	3·7	4·1	3·8
Carbohydrate	6·9	7·2	4·7
Minerals	0·2	0·2	0·6
Water	90·2	87·1	87·6
Kilocalories per 100g	67	69	65
(Kilojoules) per 100g)	(281)	(289)	(272)

Infant Formula

It is still the practice for most complete infant formulas to be based on cows' milk, but the composition is modified to be nearer that of breast milk. Modifications can be made in the following ways:

1. The proportion of total protein is reduced. In some cases the protein consists of total milk protein and additional whey protein.
2. The proportion of carbohydrate is increased, usually by increasing the lactose or sugar content.
3. The proportion of certain minerals, such as sodium may be reduced.
4. The type of fat may be changed so that the fatty acid composition is nearer to that of fat in breast milk. Some products are based entirely on vegetable fats, some contain a mixture of milk and vegetable fats, whilst some contain all milk fat.
5. Calculated quantities of vitamins and minerals are added.

Infant formulas are prepared by specialised manufacturers as liquids and as powders.

The sterilized liquid 'ready to feeds' are most often used in hospitals and are prepared in glass containers and need only warming before use. A sterilized teat unit is usually supplied with each container.

The dried preparations are reconstituted with boiled (but not boiling) water, according to instructions supplied.

The choice of an artificial food would be guided by medical advice in relation to individual needs. Advantages and disadvantages in relation to price, availability, storage and methods of preparation may also be considered.

Cows' Milk

Unmodified cows' milk can be introduced to the infant's diet at about nine months, but it is still advisable to boil before use. Cows' milk or evaporated milk is an important ingredient in many recipes incorporating other foods during the weaning of infants.

Chapter VIII

MILK AND HUMAN DISEASES

Milk-borne diseases were once a serious hazard in this country, but their significance has declined rapidly and dramatically since the late 1930s. The diseases spread by milk have included tuberculosis, brucellosis (undulant fever), scarlet fever and septic sore throat, diptheria, typhoid and para-typhoid fever, dysentery, gastro-enteritis and possibly infectious hepatitis. Excluding tuberculosis and brucellosis, the majority of cases have been caused by infection of the milk through a careless milk handler or contaminated water or utensils during one of the processes.

Outbreaks of Milk-Borne Diseases

When the outbreak of a disease is caused by milk the onset may be explosive. The first cases will probably occur in the same house or establishment where several people have drunk the contaminated milk. Usually only one source of supply is involved and when this is stopped the outbreak ceases. Very occasionally the disease may be found in foods made from the infected milk.

Tuberculosis and Brucellosis

Tuberculosis and brucellosis (undulant fever) may be transmitted by the milker but the milk is usually contaminated by direct transmission of the organisms from a cow with tuberculosis of the udder or suffering from abortus fever.

The risk of tuberculosis from milk is now almost non-existent in the United Kingdom (see page 12) and evidence exists to indicate that no cases of undulant fever are caused by drinking heat treated milk. Some are caused by drinking raw (untreated) milk while a good proportion of the comparatively few cases in this country are transmitted by direct contact through an abrasion in the skin — the more usual sufferers being veterinary officers, farmers or others who assist an infected cow when she is calving.

Improvements in the United Kingdom

Milk-borne diseases have been reduced to a very low level in the United Kingdom mainly because the heat treatment of milk has reached a level of 96 per cent of all milk consumed.

In addition bovine tuberculosis has been eradicated and a brucellosis eradication programme started in 1972. It is hoped that the latter disease will be eliminated by the early 1980s. Meanwhile, milk which is not heat treated may be sold for human consumption only if it comes from Brucellosis Accredited Herds.

At the same time, methods of improving milk quality, raising standards of hygiene on the farms and in the dairies, and eradicating bovine diseases are the subjects of constant research.

Allergies

Antibiotic Allergies and Sensitization

Residues of antibiotics and other inhibitory substances in milk are due principally to the use of such compounds for the intramammary treatment of cows suffering from mastitis, although they may also be derived from other forms of treatment. These residues could sometimes have unfortunate results if consumed by people who are allergic to antibiotics* but the number of confirmed cases is very small. Special tests are carried out to check the presence of antibiotics in milk (see page 24).

Milk Allergy

A few people are allergic to milk, but the condition normally occurs in infancy and is often of a temporary nature. However, some children are born with a deficiency of enzymes needed to convert the galactose found in milk sugar (lactose) to glucose. Hence galactose cannot be metabolised and is excreted in the urine giving rise to galactosaemia. Special dairy products that have had the lactose replaced by a more digestible sugar are used under medical supervision for the treatment of this disorder.

*See Report on Antibiotics in Milk in Great Britain 1963 (HMSO)

Chapter IX

CREAM AND BUTTER PRODUCTION

Cream is the lighter weight portion of milk and contains all the main constituents of milk, but in different proportions. The butterfat content of cream is higher than in milk and the solids-not-fat content is lower.

Butter is a substance obtained from cream by churning. Whereas milk is an emulsion of globules of fat in a watery solution, butter is an emulsion of droplets of water in fat. Commercially-made butter in the United Kingdom contains at least 82 per cent butterfat.

Butterfat

The butterfat in milk is present in the form of small globules, which can only be seen under the microscope, each one of which is surrounded by a layer made up of some of the other solid components of the milk. These globules sometimes stick together in groups or aggregates and rise to the surface to form a cream line or layer.

The manufacture of both butter and cream depend on the separation of this layer from the rest of the milk.

Creamery Laboratory Tests

On arrival at the creamery the milk is tested for hygienic quality, butterfat content and total solids (see pages 22 to 23) and, if the results are satisfactory, it is pumped into large storage tanks until required.

Separation of Cream

Cream may be separated in a variety of ways, the simplest being to allow the milk to stand for about 24 hours before skimming the cream from the surface. This method is less efficient than mechanical separation because, when skimming is employed, the cream tends to hold too much of the skim milk.

Mechanical Separators

The milk is first heated in a heat exchanger to 35°C—54°C. It is then fed to the separator. The factory separator (see diagram on page 66) consists of a stainless steel bowl enclosed in a fixed outer metal casing and rotating at about 6,000 revolutions per minute. Inside the bowl is a stack of conical plates with small spaces between each.

The milk enters the stack of plates through holes and because the skim milk is heavier than cream it is thrown towards the outside edge of the plates during rotation at high speed whilst the cream flows towards the centre. The cream and the skim milk are collected from two different outlet pipes and by means of a back pressure control valve the fat content of the cream may be adjusted to any pre-selected fat percentage required. The cream from the separator is cooled to about 4·5°C and put into large storage tanks.

Skim Milk

The skim milk, depending on its subsequent use, is either cooled immediately to 4·5°C or flash heated to a temperature of about 79·5°C to kill off any harmful bacteria and then immediately cooled to about 4·5°C; it can be used for other milk products or for feeding to animals. For economy the hot skim milk is sometimes used in the heat exchanger to heat by regeneration the cold milk coming in to achieve the required separation temperature.

Heat Treatment of cream

The exact temperature used for heat treatment varies according to the characteristics of the cream being processed, and the use to which it will be put, but it is always sufficient to kill any pathogenic bacteria present.

Cream for Consumption as Fresh Cream

Cream which is to be sold as fresh cream is pasteurized in a heat exchanger, firstly to kill harmful bacteria and secondly to improve its keeping quality without affecting the flavour. The temperature to which the cream is heated, about 85°C flash or 74°C for 15 seconds, and then cooled to 4·5°C, is controlled automatically, and after receiving this treatment the cream is pumped into tanks ready for packaging.

Cream for Buttermaking

Cream used for buttermaking usually contains 35—40 per cent butterfat. It is heat treated in a pasteurizer of the same type as that used for other cream, but the plant may also include special sections designed to remove undesirable flavour and air. In this type of plant, cream is first heated to a temperature of about 60°C in a heat exchanger. Steam is then injected into it, raising its temperature to about 79·5°C—82°C. After this the cream is sprayed into a chamber under vacuum where it boils — so releasing the added steam together with the volatile 'off-flavours' and dissolved air. In some creameries this process may be repeated before the cream is cooled in the heat exchanger to a temperature of about 4·5°C—7°C ready for holding overnight (see page 41).

A special cultured butter, commonly known as lactic butter, can also be made by addition of a "starter" of selected acid- and flavour-producing bacteria to the pasteurized cream. The milk is then incubated until the desired acidity is reached before buttermaking.

Transportation in Bulk

After cream is produced and separated in the way described, in some instances, it is transported in bulk usually in road tankers (after pasteurization) to urban areas from the country and, on arrival at the urban depot, is re-pasteurized so that any bacteria which may have entered it during transport are killed before the cream is packaged.

Filling and Capping

Machines automatically fill cream into sealed containers of a variety of designs. Each container is printed with the type and weight or volume of cream that it contains to meet legal requirements. Most companies operate a voluntary coding system for consumer protection.

Legal Control and Types of Cream

By law strict standards must be maintained in handling cream and the minimum fat contents are laid down in the Cream Regulations 1970. Under these regulations *double cream* must contain no less than 48% butterfat, *single cream* must contain no less than 18% butterfat and *half cream* must contain no less than 12% butterfat. Because of the lower fat content of single cream and half-cream they are usually homogenized to prevent the cream separating on storage and, in the case of the former, to induce greater viscosity.

Whipping cream and *whipped cream* must contain no less than 35% butterfat. *Clotted cream* is a type of cream produced mainly in the south-west and is usually sold under the name of Devonshire or Cornish clotted cream. It must have a minimum fat content of 55%. It is made in the following manner: milk is placed in shallow pans and left until the cream has risen. It is then heated to a temperature of about 82°C and allowed to cool overnight. The cream is then skimmed off. This is the traditional method. In some factories the direct scald method is employed. This involves scalding cream in shallow pans and then transferring the content into tins or bottles.

Sterilized cream is a special type of cream which will keep virtually indefinitely until opened. It is made by filling bottles or cans with homogenised cream, sealing the containers and heat treating the cream in its container, usually to a temperature of about 116°C for about 20 minutes. This intense heat treatment kills all the bacteria. It also gives the cream a different flavour from fresh cream. Sterilized cream should contain no less than 23% butterfat. A slightly thinner sterilized cream is sterilized half cream. This should contain no less than 12% butterfat.

There are several varieties of *long-keeping cream* on the market. Usually sold under brand names, the keeping quality of this type of cream is longer than that of fresh cream, but shorter than that of sterilized cream. This type of cream is pasteurized (in the glass bottles or jars in which it is distributed) at temperatures between that used for fresh cream and that used for sterilized cream. Normally this is 65·5°C for 30 minutes followed by cooling to 4·5°C. Higher temperatures with shorter holding times are sometimes used, but these tend to affect the flavour of the cream. Single cream, double cream and clotted cream may be treated in this way. *UHT cream* is also available. This is treated by a process similar to that used for UHT milk (see page 19) and it has a long shelf-life.

Composition of Cream

Nutrient	Grams per 100g	
	single cream	double cream
Protein	2·4	1·5
Fat	21·2	48·2
Carbohydrate	3·2	2·0
Minerals	0·4	0·2
Water	71·9	48·6
Kilocalories per 100g	195	447
(Kilojoules per 100g)	(876)	(1841)

Cream contains not only fat but also water, some protein, milk sugar, minerals and vitamins.

Uses of Cream

Cream is a very good energy food which may be used in a wide variety of dishes — with fresh or tinned fruit, cereals, coffee, soups, sauces, gravies, stews, blanquettes, mousses and 'creams'. Because it is easily digested, cream is very useful in feeding invalids.

The ideal *cream for whipping* has a butterfat content of between 35 per cent and 42 per cent. It should be kept for about 24 hours at a temperature of 4·5°C before whipping which may be done in various ways — manually or electrically — but, whichever method is used, care must be taken not to over-whip as this will result in a buttery product. Single cream (with a butterfat content below 35 per cent) will not normally whip unless some agent, such as egg-white, is added.

Care of Cream

Real dairy cream requires the same care in handling as milk. *Pasteurized cream* will keep for about 3—4 days in summer and 6—7 days in winter depending on the temperature of storage. *Sterilized, long-keeping cream* and *UHT cream* will keep for long periods without refrigeration as long as the container remains unopened. *Frozen cream* can be kept almost indefinitely in a deep-freeze unit, but is usually reserved for catering purposes only.

All types of cream will keep only as long as pasteurized cream once the container is opened and then it should be stored in a refrigerator.

Buttermaking

Holding of Cream

A preliminary holding period is important in determining the right condition and temperature of cream used for churning. Well-insulated holding tanks, glass-lined or of stainless steel, are usually fitted with mechanical agitators to give slight agitation. It is essential to hold cream for a sufficiently long period at a temperature of about 4·5°C to ensure uniform hardening of fat globules. The temperature at which the cream is stored and the time allowed for acid production depends on the flavour which is required. Conditions vary from 10°C—13°C overnight to 15·5°C—18·5°C for 3 or 4 hours. After this the cream is cooled to 7°C prior to churning. Holding for several hours at this temperature ensures that the fat is in the correct condition for churning, and minimises the fat losses in the buttermilk.

Cream is stored during this period in large 2,250—4,500 litre stainless steel insulated storage tanks, some of which may be fitted with refrigeration coils to prevent the temperature of the cream rising.

Churning

Traditionally butter churns have been used for converting cream into butter but these are now being replaced commercially by continuous buttermaking machines which undertake all the processes — churning, washing, salting and working — previously done by different machines. The cream is fed by pipe from the holding tanks into the buttermaker.

Principles of churning

Here the cream is churned. The effect of churning is to break the envelope of solids-not-fat around each of the small fat globules in the cream so that they can coalesce and form larger 'masses' of butterfat. The envelope is then dispersed in the liquid portion of the cream which becomes the buttermilk. It is essential that, during churning, the temperature of the cream is carefully controlled at about $7°C$ or the 'break' may be delayed or even prevented.

Washing and salting

The buttermilk is then drained off and replaced with chilled water which washes the butter grains, removing any residual buttermilk which, if left, might reduce the keeping quality of the final product. It also facilitates the hardening of the fat granules.

The buttermilk, a by-product of buttermaking, is not, however, wasted. It may be used for livestock feeding or reseparated for mixing with whole or skim milk for powdered milk manufacture.

The excess wash water is then drained off and salt, in the form of a brine solution, is added, but this is largely determined by market requirements and whether a long transport and storage period is required. Salt helps to prolong the life of butter. The percentage of salt added must be very carefully controlled and it must be worked evenly into the butter, otherwise the flavour, texture and appearance will be affected. The amount of salt added depends on the consumers' tastes. These vary greatly, but an average percentage is about 1 per cent for lactic butter and 1·5 per cent for sweet butter.

Working

The butter is then 'worked' to ensure correct consistency and moisture content according to legal requirements, which in most countries, including the United Kingdom is 16 per cent. The finished butter is discharged as a continuous ribbon from the end nozzle of the buttermaker, ready to go to the packaging machine.

Packaging

Butter is packed either in large boxes or casks or in quarter, half or one-pound packets ready for sale. Mechanical packing slightly compresses the butter, thus avoiding discolouration and deterioration due to pockets of air.

Legal Control and Grading of Butter

The composition of butter is strictly controlled in this country by the Butter Regulations 1966.

At the present time it is not compulsory for butter destined for sale in the United Kingdom to be graded, although grading facilities are offered by the National Association of Creamery Proprietors and Wholesale Dairymen. The latter awards grades under a system which includes 50 points for flavour, 20 for body and texture, 20 for colour and appearance and 10 for the absence of free moisture. Some manufacturers do not use this grading service, preferring to grade their butters themselves. Here the butters are checked organoleptically and tested chemically and bacteriologically.

Exporting countries produce butter to standards which are accepted in the United Kingdom. Imported butter sold in the United Kingdom has to comply with British legal requirements as regards composition and labelling. Butters produced in the United Kingdom but destined for EEC trade are deposited in stores run by the Intervention Board for Agricultural Produce. This body also applies grading standards similar to the National Association of Creamery Proprietors and Wholesale Dairymen, but in addition applies chemical and bacteriological standards, as well as monitoring packaging quality.

Composition of Butter

Butter usually consists of rather more than 80 per cent butterfat. It also contains vitamins A and D (the amount varying with the season) and very small amounts of protein, milk sugar and minerals. The mineral content in *salt butter* is higher than in *unsalted butter* owing to the addition of salt in the manufacture of the former.

Nutrient	Grams per 100 g salt butter
Protein	0·4
Fat	82·0
Carbohydrate	trace
Minerals	2·3
Water	15·4
Kilocalories per 100 g (Kilojoules per 100 g)	740 (3041)

The *flavour and colour of butter* vary according to several factors including:
(a) the type of cattle and/or pasture land from which the milk used for its manufacture was produced;
(b) the method of manufacture, i.e., whether it is made from fresh or ripened cream;
(c) the amount of colouring matter or salt added.

Uses of Butter

Butter is the foundation of French *haute cuisine*. It may be used wherever fat is required, in all kinds of cookery, as a garnish for meat, fish and vegetables, and as a spread.

In the home, any kind of butter is suitable for shallow frying, but, in high-temperature frying, a teaspoonful of olive oil should be added. Unsalted butter is the best to use for sauces, creams, garnishes, butter icing and brandy butter.

Care of Butter

Butter should be kept cool and away from any food with strong flavours or smells. Like milk, it is affected by light and should be stored in a dark place.

Chapter X

CHEESE PRODUCTION

Cheese is the natural and oldest way of preserving the nutrients of milk. English cheeses were traditionally made in country farmhouses. Before the old crafts were forgotten, the nine famous regional cheeses were successfully reproduced on a large scale in creameries (cheese-making factories). English cheeses today have all the flavour produced by the craftsmen and in addition a high standard of hygienic production. Comparatively slight variations during manufacture produce different varieties of cheese (see page 50).

How Traditional English Cheddar Cheese is made

Cheesemaking starts with fresh milk direct from the farm which is tested on arrival at the creamery (see page 22). If satisfactory it is heat treated to destroy any undesirable organisms which might be present. Heat treatment involves raising the temperature of the milk to between 69°C and 72°C for at least 15 seconds. The milk is then cooled to 30°C before being pumped into cheese vats which may hold up to 9,000 litres, although most modern creameries use vats of up to 22,500 litres. Here the milk is ripened.

Ripening of Milk

Ripening is initiated by adding starter in the proportion of 10 litres per 500 litres of milk (i.e. 20 per cent). Since heat treatment destroys some beneficial micro-organisms including the naturally present lactic acid-producing bacteria, it is necessary to re-introduce these as laboratory-grown cultures of bacteria which are known to convert milk sugar (lactose) into lactic acid. These are referred to as 'starter' cultures. Lactic acid acts as a preservative, thus contributing to the long-keeping quality or 'shelf life' of the cheese.

Renneting

After about half an hour, when the developed acidity in the ripened milk reaches 0·02 per cent lactic acid the milk is heated to 30°C and rennet is added. Rennet is a natural extract from the dried stomach of a calf containing the enzyme rennin which has the property of causing the milk to clot. In recent years a number of rennet-type enzymes have been developed, most of these being derived from either fungal or bacterial fermentations. For Cheddar cheese, rennet is added at the rate of 100—120 ml per 500 litres of milk.

From now on the progress of the cheesemaking process is measured by testing for the production of acid in the milk and subsequently in the whey. Milk possesses a 'natural acidity' varying between 0·13 and 0·19 per cent, but any increase in acidity is due to the activity of the starter which produces lactic acid, the increase of acidity over the original being referred to as 'developed acidity'.

Measuring the acidity

Samples of milk are tested for lactic acid content which is measured in one of three ways:

1. *The Rennet Test:* the acidity of the milk is indicated by the speed of curdling after the addition of rennet, e.g. high acid — quicker curdling, low acid — slower curdling. This test is rarely used now but has been replaced by:
2. *The Acidimeter Test:* the amount of standard solution of caustic soda needed to neutralise the acidity of the milk or whey measures the acidity directly in chemical terms. The more acid — the more solution needed.
3. *Hot Iron Test:* a piece of curd pressed against a hot iron and then slowly withdrawn produces threads of 'plastic' cheese. The length of these threads indicates the maturity of ripeness of the curd and can be applied only to the cheddaring curd. Long thread — high acidity.
   Acidity at the time of adding starter: 0·16 per cent lactic acid.
   Acidity at the time of renneting: 0·18—0·19 per cent lactic acid.

Cutting the curd

After the rennet has reacted for about 30—45 minutes the curd produced will be firm or 'solid' enough to cut into small pieces. The process of cutting releases whey from the curd. Two 'American' knives, horizontal and vertical-bladed, are passed through the curd to cut it into pea-sized cubes. 'American' knives are so-called because they were first introduced in the United States where cheese factories originated in the latter half of the nineteenth century. Alternatively, a frame strung with wire spaced at about $\frac{3}{8}$in. (9·5mm) intervals, may be used for cutting the curd.

Many modern creameries have now mounted these knives or frames on overhead movable motors and these cut the curd mechanically. When the curd has been cut, the whey produced is tested for acidity. Since one of the milk proteins, casein, and the phosphates, are now enclosed in the curd, the acidity of the whey is lower than that of the original milk and varies from 0·12—0·16 per cent lactic acid.

Scalding

The curd is then stirred continuously during the scalding period of 40—45 minutes. Stirring is done by a centrally-mounted bank of paddles which swing from side to side across the vat or by revolving overhead stirrers. Scalding involves the passing of steam or hot water through the hollow jacket surrounding the vat — thus heating the contents to 36°C where acid development has been slow, or up to 41°C where acid development has been faster. This scalding or cooking process helps to expel the whey from the curd and to obtain the desired texture. If the acidity is developing slowly, the curd should be scalded more slowly and to a lower temperature, i.e. the temperature should be raised not less than ½°C over three minutes. This helps to retain the whey which is then available to assist further acid development. If, however, the acidity is rising quickly, the temperature may be raised ½°C over two minutes or less. *Acidity before scalding:* 0·16—0·18 per cent lactic acid.

Note: Acidity is less at this stage because the milk protein (casein) which is acid, is no longer in solution. It has coagulated.

If the curd has not achieved the desired change of texture at the end of the scalding period, stirring may be continued after the heat is turned off. But with a normal or rapid development of acidity, the curd is likely to be ready for 'pitching' — it will be about wheatgrain size and will remain free and feel firm and springy when squeezed lightly in the hand. This is where the cheese-maker's skill and experience play their part, allowing him to judge the right moment to allow the curd to pitch.

Pitching (settling)

When the stirring stops after scalding, the curd is allowed to 'pitch' or settle to the bottom of the vat. Whey is run off when the developed acidity is between $0 \cdot 21 — 0 \cdot 22$ per cent from whey obtained by squeezing a handful of curd. This may take place in the vat used for the earlier processes, or the curds and whey may be piped into shallow trays called curd coolers after scalding. Here the curd is allowed to settle and the whey is run off through a central grid which is provided to assist the drainage of the curd.

Cheddaring

When the whey has been drained off, the settled curd is cut into square blocks. These blocks are piled on each side of the vat and re-piled at intervals of one to ten minutes to ensure complete draining. During the draining or cheddaring period, the curd shrinks and consolidates further. The acidity increases considerably, the curd becomes silky in texture and can be torn into leaves. The blocks are re-piled higher and higher (usually six high) during cheddaring until finally all the blocks are in piles along the edges of the vat. Consequent pressure effectively squeezes out the whey left in the curd. *Acidity at the end of the cheddaring period:* $0 \cdot 65 — 0 \cdot 75$ per cent lactic acid $3 \cdot 75$cm ($1\frac{1}{2}$ inches) by Hot Iron test).

Milling

In some creameries the blocks of curd are passed on a conveyor belt into an electrically driven mill. This cuts the curd cleanly and evenly into 'chips'. In other creameries the curd mill may be mounted on a trolley which can be wheeled over the vat. An alternative type of mill, a 'peg' mill, is sometimes used. This tears the curd into uneven pieces.

Salting

Salt is added to the milled curd at approximately two per cent. This may be done automatically or, where milling takes place in a vat, it may be mixed in with an ordinary steel-pronged fork. The salt helps to preserve the finished cheese and brings out its flavour.

Packing into moulds

The salted curd is packed into moulds lined with a coarse cloth or tubular bandage. The latter stays on the cheese throughout the pressing and ripening process. A traditional Cheddar cheese mould is a metal cylinder, with removable top and bottom discs, which contains about 30—32kg cheese. Smaller cheeses are now made and square or rectangular moulds are used for cheese which is to be pre-packed.

Pressing

The cheese in the moulds is given a pre-press for a short time only to drive out the remaining whey. The moulds are then put into a powerful horizontal press, sprayed with hot water and left for 24 hours under a pressure of 2—5 tonnes. The hot water sprayed on to the mould produces a thin hard rind on the cheese. This is necessary if the cheese is to keep properly.

Ripening

The pressed cheese is date-stamped and taken to the ripening room. This room is kept at a temperature of about 10°C and a relative humidity of 80—90 per cent. The even temperature ensures uniform ripening, and the correct humidity (high moisture content of air) prevents shrinkage of the cheese due to evaporation. The cheeses are turned top-to-bottom — at first every day, less frequently later.

During the ripening period, a change in the composition of cheese takes place. This is due to the continued action of various bacterial organisms and enzymes. In particular the casein, or milk protein, in the cheese becomes more digestible. At the same time, the product develops the characteristically rich, nutty flavour of a mature cheese. The cheese becomes firm and the rind more pronounced.

Recent advances

In recent years many advances have been made in the hard cheese-making process. So much so, that nowadays large quantities of cheese are currently manufactured in the United Kingdom on continuous mechanical systems. These processes are allied with large cheese vats and a culture preparation installation, the total systems being monitored by process control computers on pre-determined programs.

Cheddaring

The curds and whey are separated on a moving perforated belt equipped with stirring gear. At the end of this conveyor, the curd is blown to the top of a stainless steel cheddaring tower. Here the curd is held for approximately $1\frac{1}{2}$ hours whilst it drains and fuses to achieve the characteristic Cheddar cheese texture. When the curd reaches the bottom of the tower, it passes to the mill which cuts the curd into chips. It then feeds the curd to a conveyor where salt is automatically added. This is the basis of the Cheddarmaster system of cheese-making. In other systems, for example, the Bell-Siro system, the curd is cheddared in large continuously moving buckets instead of the tower system.

Pressing

Traditional pressing methods are frequently used in conjunction with these modern processes, but more modern pressing methods are also in use e.g. the "Tonne Press" which presses approximately one tonne of curd in large stainless steel boxes with perforated linings. Hydraulic rams on the floor below push upwards applying pressure to the cheeses, for about 16 hours, usually overnight. The cheese is then cut into 18kg blocks and conveyed to the wrapping room where it is placed in a plastic film bag which is then evacuated of air and hermetically sealed and shrunk. The wrapped blocks are weighed and put into cartons or wooden containers for transfer to insulated cheese stores.

Here the cheese is kept at a maximum of 10°C. If specially matured cheese is required, storage could be as long as 12 months but for mild Cheddar the average storage time is five months.

Legal Control and grading of Cheese

The Cheese Regulations 1970, prescribe fat and moisture standards and labelling requirements for all types of cheese (including processed cheese and cheese spread) and limit the substances which may be added to cheese.

Quality Control

Creamery cheese manufacturers undertake to uphold quality standards for cheese made from full cream milk as laid down by the Cheese Grading Organisation of the National Association of Creamery Proprietors and Wholesale Dairymen. Monitoring is carried out at regular intervals on the manufacturer's premises by skilled Technical Officers of the Grading Organisation to ensure compliance with quality standards. Such inspections can take place at any time but are usually as near as practicable to despatch of cheese to the market.

Grading

The NACEPE Grading Organisation also undertakes to grade cheese for manufacturers requiring such service. All cheese graded is allocated marks out of 100: 45 awarded for flavour and aroma, 40 for texture, 10 for appearance and 5 for colour. There are three main grades — *Extra Selected* when the cheese is of superlative quality; *Selected* when it is of good quality and *Second Grade* when it has not quite attained the high marks necessary for the other two grades. In the case of farmhouse cheese, the Milk Marketing Boards are responsible for grading Cheddar, Cheshire and Lancashire made on farms under contract. There are four grades — *Superfine, Fine, Graded* and *No Grade*, of which only the Superfine and Fine grades are sold as 'Farmhouse' cheese; Graded and No Grade cheese is sold for processing. Special arrangements are made where other types of cheeses are produced.

Composition of English Cheeses

The composition of cheese varies according to the milk or cream from which it is made for cheese is virtually solid milk without lactose (which remains in the whey). All the well known English cheeses are made from whole milk.

	Grams per 100g	
Nutrient	Cheddar cheese	cottage cheese
Protein	26·0	13·6
Fat	33·5	4·0
Carbohydrate	trace	1·4
Minerals	3·4	1·4
Water	37·0	78·8
Kilocalories per 100g (Kilojoules per 100g)	406 (1682)	96 (402)

Nutritional Value of Cheese

The importance of cheese in the diet lies in its contribution to the intake of protein, calcium, vitamin A and riboflavin, because it retains these valuable constituents of milk. The calorific value is high and more than 90 per cent of cheese is digestible material which may be used for growth and repair of body tissues or energy.

Uses of Cheese

Cheese is one of the most versatile and economical foods and offers considerable scope for variety in the diet. It may be eaten raw or cooked as part of a main meal or as a snack at any time of the day and can be served to almost any age-group.

Older people often prefer more mature flavours while milder varieties are usually more suited to younger tastes. From as early as nine months, babies can eat and enjoy mild cheese if it is grated.

The main varieties of English cheese are — *mild:* Caerphilly, English Cheshire, Wensleydale, mild English Cheddar, — *medium:* Lancashire, Leicester, Derby, White Stilton, — *mature:* Double Gloucester, mature English Cheddar, Blue Stilton.

Care of Cheese

Whenever possible cheese should be eaten fresh although, if properly stored, it will retain its flavour and moisture for considerably longer than most other perishable foods. Cheese should be wrapped in a polythene bag to prevent drying and then stored in a cool larder $10°C—15·5°C$ or in a refrigerator. In the latter case, the cheese should be wrapped very tightly and taken out about half an hour before serving in order to bring it to room temperature.

Acid Curd Cheeses

Acid curd cheeses are made from whole, partially skimmed or skimmed milk and the texture of the finished cheese is dependent upon the fat content of the milk. Coagulation of the curd is brought about by natural souring (the action of lactic acid bacteria which ferment the lactose to produce lactic acid) and this process may be initiated by the use of a starter which can also be used to control fermentation.

Acid curd cheeses have a short shelf-life and should be consumed fairly soon after manufacture. Moulds and yeasts grow on the surface within a few days and the flavour becomes unpleasant.

The nutritional value of these cheeses is high for they contain most of the nutrients of the milk from which they are made.

Owing to their easy digestibility they are well suited to the feeding of children and invalids.

Cottage Cheese

Cottage cheese is a creamy acid curd cheese with a distinctive, delicate flavour. It is made from pasteurized fat-free milk inoculated with a special starter to develop texture and flavour. The resultant curd is cut into small cubes and slowly heated, to develop the right body and texture. The whey is then drained off and the curd is washed and cooled. Cream and salt are then

Cheese	Notes on Manufacture	Traditional shape and size	Remarks
HARD PRESSED AND SCALDED CHEESE			
English Cheshire	Lower scalding — to only 32°C — 35°C — avoids too hard a curd. Cheshire ripens in four weeks. A red colour is sometimes introduced by adding a harmless vegetable colouring (anatto).	33 cm high 28 cm diameter	Loose and crumbly in texture. Its keen, tangy flavour is often said to be due to the salty soil in Cheshire on which the cattle graze. Cheshire cheese can be red, white or blue.
Derby	The curd is cut into 1cm cubes and is still quite moist at milling. It ripens in six months.	13cm high 36cm diameter	A beautiful white cheese with a smooth texture and soft, mild flavour. The rare Sage Derby gets extra flavour by being impregnated with the extract pressed out of finely-chopped sage leaves.
Double Gloucester	The manufacture of Double Gloucester is very close to Cheddar. Double Gloucester ripens in four to six months.	9cm high 35cm diameter	Close and smooth in texture, straw-coloured or light red. It has a mellow, fairly pungent flavour.
Leicester	The curd in the vat is drained under pressure from weighted wooden racks. Very finely milled. The cheese ripens in two months.	9cm high 35cm diameter	Leicester Cheese is a rich red colour. Mild to taste. Soft, crumbly and flaky in texture. It is shaped like a millstone.
LIGHTLY PRESSED AND SCALDED CHEESE			
Caerphilly	The new cheese is soaked in brine for 24 hours after a day's light pressing. Then dried and stored. It ripens in fourteen days.	20cm high 33cm diameter	A creamy white cheese with a mild delicate flavour and semi-smooth texture.
Lancashire	Part of the previous day's curd is mixed with part of the current day's. This produces a cheese with a loose, soft body. The curd is drained in cloth bundles on wooden racks and milled as fine as chopped suet.	20cm high 33cm diameter	Crumbly and fairly soft, with a clean and mild flavour. It is especially famous for its toasting qualities.
White Wensleydale	Only about 500ml of starter is used to 500 litres of milk. The milled curd is packed into unlined metal moulds and left to drain for 1 to 2 hours before gentle pressing. It ripens in twelve to fourteen days.	15cm high 20cm diameter	This celebrated cheese was first produced in the Middle Ages by monks of Jervaulx Abbey on the River Ure. The lingering creamy sweet-tasting flavour of Wensleydale is unique.
BLUE VEINED CHEESE	The blue-green veining in Blue Stilton and Blue Wensleydale is a harmless mould which grows in air spaces in the loosely packed curd. This is encouraged by the circulation of air from outside. The growth of this mould is essential to develop the clean, piquant flavour of a blue cheese.		
Blue Stilton	The curd is cut; ladled into shallow sinks lined with perforated trays; left over-night. It is then broken very carefully by hand into small pieces. Salted and packed gently into hoops. The hoops are drained on calico squares laid on boards; the cloths and boards being changed daily for 10 days. Then the hoop is removed. The sides of the cheese scraped, and the scrapings used to fill the crevices which forms the basis for a 'coat'. As the 'coat' forms a white mould begins to grow on the surface. Blue veins form in cracks in the cheese body. To allow them to form properly the cheese must ripen very slowly. Aeration encourages the growth of this mould and may be increased by piercing the cheese with wires.	23cm high 20cm diameter	Stilton is sometimes called the 'king of cheeses.' It has a wrinkled brown coat, a creamy-white, blue-veined body. Its taste is superb — rich and 'edgy'.
Blue Wensleydale	Similar to Stilton but the soft curds are drained on racks. Very light pressure is applied to the curds in the hoops.	15cm high 20cm diameter	Gets its honeyed taste from the limestone in the soil of the Yorkshire Ure Valley.

added and, when thoroughly blended, the cottage cheese is packed into hygienic cartons. Cottage cheese contains most of the nutrients in milk but is particularly valuable as a source of protein and riboflavin. It contains significant amounts of calcium and phosphorus and is easily digested. Because it does not keep well it should be stored in a cool place.

There are many ways of introducing cottage cheese into the diet in both sweet and savoury dishes. It is especially useful in the feeding of babies, invalids and old people.

Processed Cheese

Processed cheese, or process cheese as it is called in North America, is a product prepared from Cheddar and other types of natural cheese, usually the hard-pressed varieties.

It is made by breaking down the natural cheese, finely grinding it and then emulsifying it with certain salts (especially citrates and phosphates), and often water, whey powder or paste. It is heated and thoroughly mixed into a homogeneous, pliable mass. Colouring, chopped meat or fish, spices or other flavouring may be added. The mixture is then packed in foil or other packaging material in a variety of shapes and sizes — individual portions, blocks, slices or as cheese spread.

The producer may be identified with the variety of natural cheese used in its manufacture but more than one variety may be used.

Processed cheese need not be refrigerated until opened as it is practically sterile. It has a mild flavour, keeps well and as a food is nutritious and wholesome. Compared with real cheese it is expensive and in the processing loses its natural flavour and texture.

Cream Cheeses

Strictly speaking, cream cheeses are not true cheeses because they are not made from any basic curd. Cream of various 'thicknesses' may be used — the more usual containing either 60 per cent or 30 per cent butterfat.

Cream cheeses contain the same nutrients as cream but a higher percentage of fat and much lower percentage of water. Their composition is controlled by the Cheese Regulations 1970.

Chapter XI

OTHER DAIRY FOODS

Dried Milks or Milk Powders

Milk powder is produced by the evaporation of water from milk by heat, or other means, to produce solids containing 5 per cent or less moisture. The milk is homogenized, heat treated and usually pre-concentrated in a vacuum before drying.

Roller Drying

The milk is spread on hot revolving rollers. The water evaporates rapidly to leave a thin film of powder on the rollers which is automatically scraped off. This method is used for baby foods, but the resulting powder does not reconstitute very easily. There is a tendency to cling in flakes and to form lumps when mixed with water.

Spray Drying

The milk is pumped through a fine jet as a spray into a chamber through which hot air is circulated. The fine droplets of milk quickly lose their water by evaporation, and fall to the floor of the chamber as a fine powder.

The temperature of the hot air is carefully regulated so that the protein will not be coagulated and the product will not have a strong cooked flavour.

Provided the powder is finely divided it reconstitutes easily when mixed with water. Dried skimmed milk powder reconstitutes better when it has been 'instantized', a process which forms aggregates of particles which are more easily wetted and dispersed on reconstitution.

Legal Control of Milk Powders

The Condensed Milk and Dried Milk Regulations 1977 came into force on 1st July 1978 and implement the United Kingdom's obligations under EEC Directive No. 76/118/EEC on the approximation of the laws of Member States relating to certain partly, or wholly dehydrated preserved milk for human consumption. Products complying with the new regulations were allowed to be sold from 1st July 1977. The Dried Milk Regulations 1965 continue to apply to products manufactured before 1st July 1978 until they are revoked on 1st July 1980.

They set out requirements for the labelling of containers, and advertising. In particular, the label must carry a reserved description, and any dried milk product containing less than 26.0 per cent milk fat is required, on or before 31st December 1980, to carry the declaration "not to be used for babies except under medical advice". The regulations also require the milk ingredients used in the preparation of dried milk products to have been subjected to a heat treatment at least equivalent to pasteurization if the product itself is not so treated, and specify the added ingredients and their limits permitted in dried milk.

The new regulations prescribe reserved descriptions for dried milk products as follows, and restrict sales to these products:

Reserved Descriptions	Percentage Milk Fat	Percentage Water
Dried high-fat milk High-fat milk powder	not less than 42.0 but not more than 65.0	Not more than 5.0
Dried whole milk Whole milk powder	not less than 26.0 but not more than 42.0	
Dried partly skimmed milk Partly skimmed milk powder	not less than 1.5 but not more than 26.0	
Dried skimmed milk Skimmed milk powder	not more than 1.5	

Composition of Milk Powders

Dried whole milk powder contains all the nutrients of milk in a concentrated form with the exception of vitamin C, thiamin and vitamin B_{12}. (These are partially destroyed by heat during the manufacturing process—see page 32.)

Dried skimmed milk contains virtually no fat and therefore no fat-soluble vitamins, but it does contain protein, calcium and riboflavin.

Nutrient	Grams per 100g	
	whole milk powder	skimmed milk powder
Protein	26·3	36·4
Fat	26·3	1·3
Carbohydrate	39·4	52·8
Minerals	4·6	4·8
Water	2·9	4·1
Kilocalories per 100g (Kilojoules per 100g)	490 (2051)	355 (1512)

Dried milks and products based on dried milk are used extensively in the artificial feeding of infants (see pages 34 to 35), but they are also useful when milk is to be kept for a long time or where transport in bulk presents problems, for example, on sea voyages.

When packed in air-tight containers milk powders can be kept for a considerable time at a moderate temperature. If fat is present this is liable to go rancid on exposure to the air. Once reconstituted, such milks should receive the same care as fresh milk.

Evaporated and Condensed Milks

Condensed milk is now less popular than it once was. It can be made from either whole, partly skimmed or skimmed milk. The milk is homogenised, then heated to about 80°C and held at this temperature for 15 minutes. During this period cane sugar may be added and the mixture is drawn into a large metal evaporator where it is boiled under vacuum at temperatures slightly in excess of 50°C until it has reached a concentration about 2½ times that of the original milk. The condensed milk is then drawn off, cooled and aseptically sealed into tins or barrels.

Evaporated milk is a concentrated milk and the process is essentially the same as that for condensed milk except that it does not contain added sugar and it is sterilized in the tin at 115·5°C for 20 minutes. The final concentration is about twice that of the original milk.

Legal Control of Condensed Milks

Condensed milks are also controlled by the Condensed Milk and Dried Milk Regulations 1977. The Condensed Milk Regulations 1959 will, like the Dried Milk Regulations 1965, continue to apply to products manufactured before 1st July 1978 until their revocation on 1st July 1980. The reserved descriptions applied to condensed milk products are as follows:

Reserved Descriptions	Percentage Milk Fat	Percentage Total Milk Solids
Unsweetened condensed high-fat milk	not less than 15·0	not less than 26·5
Evaporated milk	not less than 9·0 but not more than 15·0	not less than 31·0
Unsweetened condensed milk	not less than 7·5 but not more than 15·0	not less than 25·0
Unsweetened condensed partly skimmed milk	(i) retail sales — not less than 4·0 but not more than 4·5 (ii) sales otherwise than by retail — not less than 1·0 but not more than 7·5	not less than 24·0 not less than 20·0
Unsweetened condensed skimmed milk	not more than 1·0	not less than 20·0
Sweetened condensed milk	(i) retail sales — not less than 9·0 (ii) sales otherwise than by retail — not less than 8·0	not less than 31·0 not less than 28·0
Sweetened condensed partly skimmed milk	(i) retail sales — not less than 4·0 but not more than 4·5 (ii) sales otherwise than by retail — not less than 1·0 but not more than 8·0	not less than 28·0 not less than 24·0
Sweetened condensed skimmed milk	not more than 1·0	not less than 24·0

Preservation is achieved by sterilization in the case of unsweetened and evaporated products and by the addition of sucrose in the case of sweetened products. The other main requirements of the regulations apply to condensed milk products in the same way as to dried milk products. The statement of unsuitability for babies is required for condensed milk products containing not more than 4·5 per cent milk fat.

Composition of Condensed Milks

Nutrient	Grams per 100g		
	condensed whole sweetened	condensed skimmed sweetened	evaporated whole unsweetened
Protein	8·3	9·9	8·6
Fat	9·0	0·3	9·0
Carbohydrate	55·5	60·0	11·3
Minerals	1·1	1·8	1·2
Water	25·8	27·0	68·6
Kilocalories per 100g	322	267	158
(Kilojoules per 100g)	(1362)	(1139)	(660)

Uses of Condensed Milks

Condensed milks have a variety of uses in both the undiluted and diluted forms, although their pronounced flavours make them unacceptable to some tastes.

Care of Condensed Milks

Such milks will keep almost indefinitely as long as the can remains unopened. Once reconstituted, they require the same care and attention as fresh milk.

Skimmed Milk

Skimmed milk, or skim, is the liquid left after the cream has been skimmed off or separated and contains the valuable protein, calcium, lactose and water-soluble vitamins. Skim has many uses, but it is not suitable for feeding infants because of the lack of fat.

Composition of Skimmed Milk

Nutrient	Grams per 100g
Protein	3·4
Fat	0·1
Carbohydrate	5·0
Minerals	0·5
Water	90·9
Kilocalories per 100g	33
(Kilojoules per 100g)	(142)

Buttermilk

Buttermilk is a liquid by-product of the manufacture of butter. It may be sweet or sour according to whether fresh or ripened cream has been used. It contains some protein, lactose and minerals, but it is seldom used in human diet in this country, although it is popular elsewhere. Sometimes it is used to make cultured buttermilk — a refreshing fermented milk drink.

Cultured and Fermented Milks

Processes of fermenting milk in order to preserve its nutrients or to improve its digestibility have been used for many centuries. The principal types of fermented, or cultured, milks are kefir, kumiss (or koumiss) and yogurt. The best known is yogurt.

Yogurt

Yogurt may be made from whole, partially skimmed, evaporated or dried milk or any mixture of these. In commercial manufacture the milk is homogenized, heat treated and then cooled to 38°C. At this stage a specially prepared culture of *Lactobacillus bulgaricus, Streptococcus thermophilus* and sometimes *Lactobacillus acidophilus* is added and the mixture incubated at 43°C in large tanks until the desired level of acidity is reached and clotting takes place. The yogurt is then cooled and, if desired, flavouring may be added before the yogurt is filled into cartons.

Yogurt contains all the nutrients of the milk from which it was made with the exception of lactose which is used up by the bacteria in making the lactic acid which ferments the milk. Some brands have added vitamins.

Yogurt is available plain or in a wide variety of flavours and often has pieces of fruit added during manufacture. It can be used in many sweet or savoury dishes. It is highly recommended for invalids and old people and is particularly good for infants who are being weaned from a liquid to a solid diet.

During storage, the organisms present in yogurt are still active, and the acidity increases until the product becomes too acid to be palatable for most people. This increase in acidity occurs more slowly if the yogurt is kept cool. If the container is shaken, plain or flavoured yogurt curd may break, releasing a small amount of whey. Although this affects the appearance of the product there is no loss of flavour or food value.

Composition of yogurt (low fat)

Nutrient	Grams per 100g		
	Natural	Flavoured	Fruit
Protein	5·0	5·0	4·8
Fat	1·0	0·9	1·0
Carbohydrate	6·2	14·0	17·9
Minerals	0·8	0·8	0·8
Water	85·7	79·0	74·9
Kilocalories per 100g	52	81	95
(Kilojoules per 100g)	(216)	(342)	(405)

Cultured Cream

Cultured cream is prepared in a similar way to that described for yogurt. It is made from cream and contains the same nutrients as cream with the exception that in cultured cream the lactose has been converted into lactic acid. It may be used as purchased or used in any recipe which calls for sour cream. Like yogurt it is excellent in both sweet and savoury dishes.

Whey

Whey is the watery liquid which separates from the curd during cheese-making. It contains protein, lactose, fat and some minerals. Whey is potentially valuable as a human food because of the high quality of its protein and the dairy industry is currently concerned to find efficient and economic methods of processing it. At the moment, however, it is little used in human nutrition.

In the creamery the small amount of butter fat is usually separated off and turned into whey butter. After separating the fat, the whey may then be condensed and dried. This whey powder is a valuable source of protein which can be used in concentrated cattle feeds.

Frozen Milk

In the preparation of frozen milk, pasteurized milk is homogenized, poured into polythene bags and quickly frozen in a brine bath. This technique makes it possible to store the milk in deep-freeze conditions for at least 12 months without deterioration.

The commercial possibilities of frozen milk include the provision of fresh milk for passengers throughout long sea voyages.

Skimmed Milk with Non-Milk Fat Products

Skimmed milk with non-milk fat is milk which has been skimmed to remove the butterfat and then non-milk fats are substituted for the latter. It is commonly known as *filled milk*, although this is a term which has no meaning in law. Dried skimmed milk with non-milk fat products are now widely available. These powders when reconstituted may be used as alternatives to natural milk. The advertising and labelling of this type of milk are controlled by the Skimmed Milk with Non-Milk Fat Regulations 1960 (as amended) but specific compositional requirements are not laid down.

Dairy Ice Cream

Dairy ice cream is a combination of milk and cream products to which sugar, emulsifying, stabilizing and flavouring materials are added. The word 'dairy' in its name indicates that the product contains no fat other than milk fat, whereas ordinary ice cream is more often made from skimmed milk and vegetable fat. The Ice Cream Regulations 1967 prescribe compositional requirements for ice cream, dairy ice cream and milk ice.

Ice cream contains appreciable amounts of the protein, calcium, phosphorus and vitamins of the B complex of milk and will provide energy from its content of sugar, fat and milk. Its use as part of a main meal is now common practice and it can make a useful contribution to the diet in terms of nutritional value.

Chocolate Crumb

Full cream milk is concentrated, normally in an evaporator. The concentrate is then mixed with sugar, added at the rate of 30 per cent by weight. After stirring, the cocoa powder (or more likely hot chocolate liquor) is blended in. After thorough mixing the heavy, sticky mass passes through a tunnel drier where it is dried by a counter current of hot air. It is then chopped into small pieces and stored until used in the manufacture of chocolate.

APPENDIX 1   **LEGISLATION**

The following list is intended as a general but not exhaustive guide to the main current legislation concerning milk and milk products. It is hoped that the geographical subdivision will be found useful, but it is an unofficial compilation which may not be accurate in every particular.

FRESH MILK (including skimmed and semi-skimmed milk).

Great Britain

Weights and Measures Act 1963 c.31

Weights and Measures (Milk and Solid Fuel Vending Machines) Regulations 1965 (S.I. 1965/2149, as amended by S.I. 1976/795)

Weights and Measures (Exemption) (Milk) Order 1966 (S.I. 1966/237)

Milk (Great Britain) Order 1971 (S.I. 1971/1038, as amended by S.I. 1974/1549, S.I. 1975/1664, S.I. 1977/2054 and S.I. 1978/469). N.B. prices legislation is changed relatively frequently

Milk (Extension of Period of Control of Maximum Prices) Order 1974 (S.I. 1974/2139)

Weights and Measures (Marking of Goods and Abbreviations of Units) Regulations 1975 (S.I. 1975/1319, as amended by S.I. 1977/1683)

Weights and Measures &c. Act 1976 c.77

Weights and Measures (Prepacked Milk in Vending Machines) Order 1976 (S.I. 1976/794)

Price Marking (Prepacked Milk in Vending Machines) Order 1976 (S.I. 1976/796)

Welfare Food Order 1977 (S.I. 1977/25, as amended by S.I. 1977/1492, S.I. 1977/1620 and S.I. 1978/269)

England and Wales

Food and Drugs Act 1955 c.16 (as amended by the European Communities Act 1972 c.68)

Milk and Dairies (Channel Islands and South Devon Milk) Regulations 1956 (S.I. 1956/919)

Milk and Dairies (General) Regulations 1959 (S.I. 1959/277, as amended by S.I. 1977/171)

Provision of Milk and Meals Regulations, 1969 (S.I. 1969/483 as amended by S.I. 1970/511, S.I. 1971/1368 and S.I. 1978/959)

Food and Drugs (Milk) Act 1970 c.3

Milk and Dairies (Semi-Skimmed and Skimmed Milk) (Heat Treatment and Labelling) Regulations 1973 (S.I. 1973/1064)

Drinking Milk Regulations 1976 (S.I. 1976/1883) to be read in conjunction with Council Regulation (EEC) No. 1411/71 (O.J. No. L148, 3.7.71, p4), as amended by Council Regulation (EEC) No. 566/76 (O.J. No. L67, 15.3.76, p23), and the guideline figure for fat content of standardised whole milk fixed annually by the Council

Education Act 1976 c.9

Milk and Dairies (Milk Bottle Caps) (Colour) Regulations 1976 (S.I. 1976/2186)

Milk (Special Designation) Regulations 1977 (S.I. 1977/1033)

Scotland

Milk and Dairies (Scotland) Acts 1914 and 1922 (1914 c.46, 1922 c.54), and local dairy byelaws

Milk Act 1934 c.51

Milk and Dairies (Scotland) Order 1934 (S.R. & O. 1934/675, as amended by S.I. 1956/2110)

Milk (Special Designations) Act 1949 c.34

Food and Drugs (Scotland) Act 1956 c.30 (as amended by the European Communities Act 1972 c.68)

Milk (Special Designations) (Scotland) Order 1965 (S.I. 1965/253, as amended by S.I. 1966/1573 and S.I. 1975/1997)

Milk (Special Designations) (Specified Areas) (Scotland) Orders 1952 to 1977, of which the last (S.I. 1977/1047) made S.D. comprehensive in Scotland

Milk and Dairies (Channel Islands and South Devon Milk) (Scotland) Regulations 1967 (S.I. 1967/81)

Food and Drugs (Milk) Act 1970 c.3

Milk and Dairies (Semi-Skimmed and Skimmed Milk) (Heat Treatment and Labelling) (Scotland) Regulations 1974 (S.I. 1974/1356)

Milk Bottles (Labelling and Cap Colour) (Scotland) Order 1976 (S.I. 1976/875)

Bulk Transport of Milk (Scotland) Order 1976 (S.I. 1976/1857)

Drinking Milk (Scotland) Regulations 1976 (S.I. 1976/1883), to be read in conjunction with Council Regulation (EEC) No. 1411/71 (O.J. No. L148, 3.7.71, p4) as amended by Council Regulation (EEC) No. 566/76 (O.J. No. L67, 15.3.76, p23), and the guideline figure for fat content of standardised whole milk fixed annually by the Council

Milk and Meals (Scotland) Regulations 1978 (S.I. 1978/969)

Northern Ireland

Milk Acts (Northern Ireland) 1950 c.31 and 1963 c.11

Food and Drugs Act (Northern Ireland) 1958 c.27

Milk Regulations (Northern Ireland) 1963 (S.R. & O 1963/44, as amended by S.R. & O 1967/8, S.R. & O 1973/30, S.R. 1974/128 and S.R. 1976/376)

Weights and Measures Act (Northern Ireland) 1967 c.6

Agriculture (Miscellaneous Provisions) Act (Northern Ireland) 1967 c.15

Weights and Measures (Milk and Solid Fuel Vending Machines) Regulations (Northern Ireland) 1967 (S.R. & O 1967/222, as amended by S.R. 1976/327)

Agriculture (Miscellaneous Provisions) Act (Northern Ireland) 1970 c.20

Milk (Northern Ireland) Order 1971 (S.I. 1971/1037, as amended by S.I. 1974/1548 and S.I. 1977/2055). N.B. prices legislation is changed relatively frequently

Weights and Measures (Marking of Goods and Abbreviations of Units) Regulations (Northern Ireland) 1976 (S.R. 1976/155)

Price Marking (Prepacked Milk in Vending Machines) Order (Northern Ireland) 1976 (S.R. 1976/304)

Weights and Measures (Prepacked Milk in Vending Machines) Order (Northern Ireland) 1976 (S.R. 1976/352)

Welfare Foods Regulations (Northern Ireland) 1977 (S.R. 1977/7, as amended by S.R. 1977/276)

Drinking Milk Regulations (Northern Ireland) 1977 (S.R. 1977/8) to be read in conjunction with Council Regulation (EEC) No. 1411/71 (O.J. No. L148, 3.7.71, p4), as amended by Council Regulation (EEC) No. 566/76 (O.J. No. L67, 15.3.76, p23), and the guideline figure for fat content of standardised whole milk fixed annually by the Council

MILK PRODUCTS — GENERAL

The following general legislation applies to all milk products, and is not mentioned further unless it has a specific application to a product.

United Kingdom

Trade Descriptions Act 1968 and 1972 (1968 c.29, 1972 c.34)

Trade Descriptions (Indication of Origin) (Exemptions No. 1) Directions 1972 (S.I. 1972/1886)

Weights and Measures &c Act 1976 c.77

Great Britain

Weights and Measures Act 1963 c.31

Weights and Measures (Marking of Goods and Abbreviations of Units) Regulations 1975 (S.I. 1975/1319, as amended by S.I. 1977/1683)

England and Wales
Food and Drugs Act 1955 c.16
Food and Drugs (Control of Food Premises) Act 1976
Labelling of Food Regulations 1970 (S.I. 1970/400, as amended by S.I. 1972/1510, S.I. 1976/859 and S.I. 1978/646)
Food Hygiene (General) Regulations 1970 (S.I. 1970/1172)
Food Hygiene (Markets, Stalls and Delivery Vehicles) Regulations 1966 (S.I. 1966/791, as amended by S.I. 1966/1487)
Antioxidants in Food Regulations 1978 (S.I. 1978/105) (applies to 'dairy product' as defined in the Regulations)

Scotland
Food and Drugs (Scotland) Act 1956 c.30
Food Hygiene (Scotland) Regulations 1959 (S.I. 1959/413, as amended by S.I. 1959/1153, S.I. 1961/622, S.I. 1966/967 and S.I. 1978/173)
Labelling of Food (Scotland) Regulations 1970 (S.I. 1970/1127, as amended by S.I. 1972/1790, S.I. 1976/1176, S.I. 1977/1027 and S.I. 1978/927)
Antioxidants in Food (Scotland) Regulations 1978 (S.I. 1978/492) (applies to 'dairy product' as defined in the Regulations)

Northern Ireland
Food and Drugs Act (Northern Ireland) 1958 c.27
Marketing of Milk Products Act (Northern Ireland) 1958 c.31
Food Hygiene Regulations (Northern Ireland) 1964 (S.R. & O 1964/129)
Marketing of Milk Products Regulations (Northern Ireland) 1966 (S.R. & O 1966/204)
Weights and Measures Act (Northern Ireland) 1967 c.6
Agriculture (Miscellaneous Provisions) Act (Northern Ireland) 1967 c.15
Agriculture (Miscellaneous Provisions) Act (Northern Ireland) 1970 c.20
Labelling of Food Regulations (Northern Ireland) 1970 (S.R. & O 1970/80, as amended by S.R. & O 1972/318 and S.R. 1976/212)
Weights and Measures (Marketing of Goods and Abbreviations of Units) Regulations 1976 (S.R. & O 1976/155)
Antioxidants in Food Regulations (Northern Ireland) 1978 (S.R. 1978/112) (applies to 'dairy product' as defined in the Regulations)

BUTTER
United Kingdom
Trade Descriptions Acts 1968 and 1972 (1968 c.29, 1972 c.34)
Trade Descriptions (Indication of Origin) (Exemptions No. 1) Directions 1972 (S.I. 1972/1886)
Butter Prices Order 1978 (S.I. 1978/97 as amended by S.I. 1978/835). N.B. prices legislation is changed relatively frequently

Great Britain
Weights and Measures Act 1963 (Edible Fats) Order 1976 (S.I. 1976/430)
Weights and Measures Act 1963 (Various goods) (Termination of Imperial Quantities) Order 1977 (S.I. 1977/2058)

England and Wales
Butter Regulations 1966 (S.I. 1966/1074)
Colouring Matter in Food Regulations 1973 (S.I. 1973/1340)

Scotland
Butter (Scotland) Regulations 1966 (S.I. 1966/1252 as amended by S.I. 1973/1310)
Colouring Matter in Food (Scotland) Regulations 1973 (S.I. 1973/1310 as amended by 1974/1340, S.I. 1975/1595 and S.I. 1976/2232)

Northern Ireland
Marketing of Milk Products Act (Northern Ireland) 1958 c.31

Marketing of Milk Products Regulations (Northern Ireland) 1966 (S.R. & O. 1966/204)
Butter Regulations (Northern Ireland) 1966 (S.R. & O. 1966/205)
Colouring Matter in Food Regulations (Northern Ireland) 1973 (S.R. & O. 1973/466)
Weights and Measures (Edible Fats) Order (Northern Ireland) 1976 (S.R. 1976/288)

CHEESE

Great Britain
Price Marking (Cheese) Order 1977 (S.I. 1977/1334)
Weights and Measures Act 1963 (Cheese) Order 1977 (S.I. 1977/1335)

England and Wales
Cheese Regulations 1970 (S.I. 1970/94, as amended by S.I. 1974/1122)
Colouring Matter in Food Regulations 1973 (S.I. 1973/1340, as amended by S.I. 1976/2086)
Emulsifiers and Stabilisers in Food Regulations 1975 (S.I. 1975/1486, as amended by S.I. 1976/1886)
Preservatives in Food Regulations 1975 (S.I. 1975/1487)

Scotland
Cheese (Scotland) Regulations 1970 (S.I. 1970/108, as amended by S.I. 1974/1337, S.I. 1975/1597 and S.I. 1976/2232)
Colouring Matter in Food (Scotland) Regulations 1973 (S.I. 1973/1310, as amended by S.I. 1974/1340, S.I. 1975/1595 and S.I. 1976/2232)
Emulsifiers and Stabilisers in Food (Scotland) Regulations 1975 (S.I. 1975/1597, as amended by S.I. 1976/1911 and S.I. 1976/914)
Preservatives in Food (Scotland) Regulations 1975 (S.I. 1975/1598, as amended by S.I. 1976/1910 and S.I. 1977/1860)

Northern Ireland
Cheese Regulations (Northern Ireland) 1970 (S.R. & O 1970/14, as amended by S.R. 1974/177)
Colouring Matter in Food Regulations (Northern Ireland) 1973 (S.R. & O. 1973/466), as amended by Colouring Matter in Food (Amendment) Regulations (Northern Ireland) 1976 (S.R. 1976/382)
Preservatives in Food Regulations (Northern Ireland) 1975 (S.R. 1975/277)
Emulsifiers and Stabilisers in Food Regulations 1975 (S.R. 1975/278, as amended by S.R. 1976/367)

CONDENSED MILK
*Regulations marked with an asterisk are applicable only to goods manufactured before 1.7.78 and sold before 1.7.80.

England and Wales
Condensed Milk and Dried Milk Regulations 1977 (S.I. 1977/928)
*Condensed Milk Regulations 1959 (S.I. 1959/1098)
Colouring Matter in Food Regulations 1973 (S.I. 1973/1340)

Scotland
Condensed Milk and Dried Milk (Scotland) Regulations 1977 (S.I. 1977/1027)
*Condensed Milk (Scotland) Regulations 1959 (S.I. 1959/1115)

Northern Ireland
Condensed Milk and Dried Milk Regulations (Northern Ireland) 1977 (S.R. 1977/196)
*Condensed Milk Regulations (Northern Ireland) 1961 (S.R. & O. 1961/31)
Colouring Matter in Food Regulations (Northern Ireland) 1973 (S.R. & O. 1973/466)

CREAM

England and Wales
Food and Drugs Act 1955 c.16
Milk and Dairies (General) Regulations 1959 (S.I. 1959/277, as amended by S.I. 1977/171)
Cream Regulations 1970 (S.I. 1970/752, as amended by the Emulsifiers and Stabilisers in Food Regulations 1975 (S.I. 1975/1486)
Preservatives in Food Regulations 1975 (S.I. 1975/1487)

Scotland
Milk and Dairies (Scotland) 1934 (S.I. 1934/675, as amended by S.I. 1956/2110)
Food and Drugs (Scotland) Act 1956 c.30
Cream (Scotland) Regulations 1970 (S.I. 1970/1191, as amended by S.I. 1975/1597)
Preservatives in Food (Scotland) Regulations 1975 (S.I. 1975/1598, as amended by S.I. 1976/1910 and S.I. 1977/860)

Northern Ireland
Food and Drugs Act (Northern Ireland) 1958 c.27
Marketing of Milk Products Act (Northern Ireland) 1958 c.31
Marketing of Milk Products Regulations (Northern Ireland) 1966 (S.R. & O 1966/204)
Agriculture (Miscellaneous Provisions) Act (Northern Ireland) 1967 c.15
Cream Regulations (Northern Ireland) 1970 (S.R. & O 1970/194) as amended by Cream (Amendment) Regulations (Northern Ireland) 1976 (S.R. 1976/15)
Preservatives in Food Regulations (Northern Ireland) 1975 (S.R. & O 1975/277)

DRIED MILK

*Regulations marked with an asterisk are only applicable to goods manufactured before 1.7.78 and sold before 1.7.80

England and Wales
Condensed Milk and Dried Milk Regulations 1977 (S.I. 1977/928)
*Dried Milk Regulations 1965 (S.I. 1965/363)
Colouring Matter in Food Regulations 1973 (S.I. 1973/1340)

Scotland
Condensed Milk and Dried Milk (Scotland) Regulations 1977 (S.I. 1977/1027)
Dried Milk (Scotland) Regulations 1965 (S.I. 1965/1007)

Northern Ireland
Condensed Milk and Dried Milk Regulations (Northern Ireland) 1977 (S.R. 1977/196)
*Dried Milk Regulations (Northern Ireland) 1965 (S.R. & O 1965/44)
Colouring Matter in Food Regulations (Northern Ireland) 1973 (S.R. & O 1973/466)

SKIMMED MILK WITH NON-MILK FAT

Great Britain
Welfare Food Order 1977 (S.I. 1977/25, as amended by S.I. 1977/1492, S.I. 1977/1620 and S.I. 1978/269)

England and Wales
Skimmed Milk with Non-Milk Fat Regulations 1960 (S.I. 1960/2331, as amended by S.I. 1976/103)

Scotland
Skimmed Milk with Non-Milk Fat (Scotland) Regulations 1960 (S.I. 1960/2437, as amended by S.I. 1976/294)

Northern Ireland
Skimmed Milk with Non-Milk Fat Regulations (Northern Ireland) 1961 (S.R. & O 1961/190, as amended by S.R. 1976/70)
Welfare Foods Regulations 1977 (Northern Ireland) S.R. 1977/7, as amended by S.R. 1977/276)

APPENDIX 2   FOR FURTHER READING

For further information the following books may be found useful:

GLYN CHRISTIAN
Cheese and Cheesemaking (MacDonald Guidelines Series)

SIR S. DAVIDSON, R. PASSMORE, J. F. BROCK & A. S. TRUSWELL
Human Nutrition and Dietetics (Churchill Livingstone, 6th Edition)

J. G. DAVIS
Cheese — Volume I & II (Churchill Livingstone)
Dictionary of Dairying — Supplementary edition (Leonard Hill)

N. G. FOWLER
Beef and Dairy Management and Production (Hutchinson Educational)

HARVEY & HILL
Milk Production & Control (H. K. Lewis)

MINISTRY OF AGRICULTURE, FISHERIES AND FOOD
Advisory Leaflet   29 — *Compositional Quality of Milk* (H.M.S.O.)
Advisory Leaflet 222 — *Cream Cheese* (H.M.S.O.)
Advisory Leaflet 416 — *Hygienic Milk Production* (H.M.S.O.)
Advisory Leaflet 422 — *The Use of Chemicals for Hand Cleansing of Farm Dairy Equipment* (H.M.S.O.)
Advisory Leaflet 437 — *Farmhouse Butter Making* (H.M.S.O.)
Advisory Leaflet 438 — *Clotted Cream* (H.M.S.O.)
Advisory Leaflet 458 — *Soft Cheese* (H.M.S.O.)
Advisory Leaflet 495 — *Cream* (H.M.S.O.)

Fixed Equipment of the Farm Leaflet 3: *Farm Dairies* (H.M.S.O.)
Fixed Equipment of the Farm Leaflet 22: *Loose Housing of Dairy Cows* (H.M.S.O.)

Manual of Nutrition (H.M.S.O. Eighth Edition, third impression)
Technical Bulletin No. 17 — *Bacteriological Techniques for Dairy Purposes* (H.M.S.O.)

S. OGILVY
Making Cheese (B. T. Batsford)

A. A. PAUL & D. A. T. SOUTHGATE
McCance and Widdowson's *Composition of Foods* (H.M.S.O. 4th Edition)

J. W. G. PORTER
Milk and Dairy Foods (Oxford University Press)

E. P. PUBLISHING
Cheeses of England (E. P. Publishing Ltd.)
Farmhouse English Cheese (E. P. Publishing Ltd.)

*Available from the English Country Cheese Council, price 60p

MAGNUS PYKE
Success in Nutrition (John Murray)

KENNETH RUSSELL
The Principles of Dairy Farming (7th Edition revised by S. Williams, Farming Press)

SOCIETY OF DAIRY TECHNOLOGY
The Pasteurising Plant Manual (The Society of Dairy Technology)
Cream Production Manual (The Society of Dairy Technology)
Quality Control of Milk Products (The Society of Dairy Technology)
Ultra-High-Temperature Processing of Dairy Products (The Society of Dairy
  Technology)

FLOW DIAGRAM OF HEAT EXCHANGER

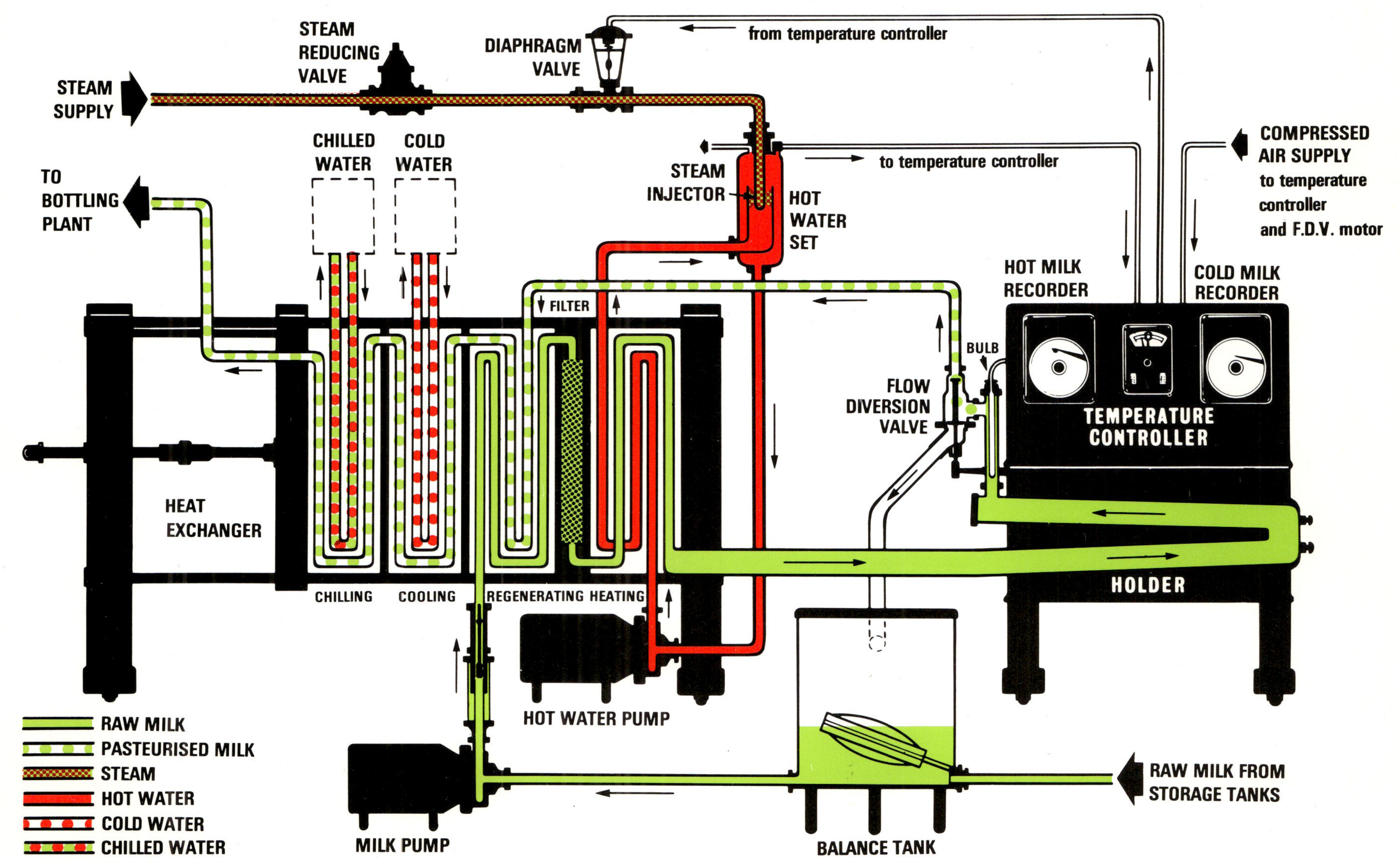

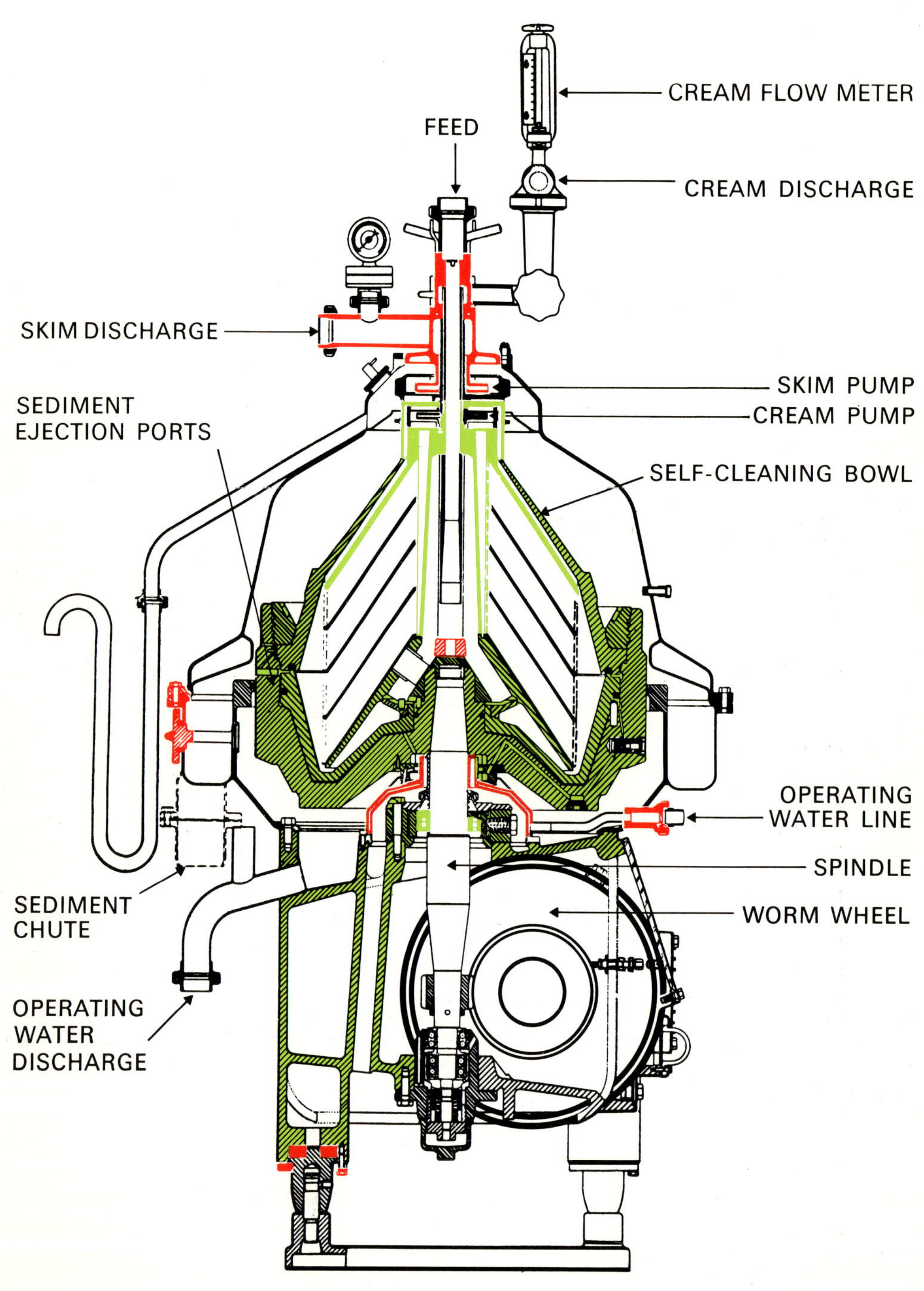
FEED
CREAM FLOW METER
CREAM DISCHARGE
SKIM DISCHARGE
SKIM PUMP
CREAM PUMP
SELF-CLEANING BOWL
SEDIMENT EJECTION PORTS
OPERATING WATER LINE
SPINDLE
WORM WHEEL
SEDIMENT CHUTE
OPERATING WATER DISCHARGE

NOTES